This book provides an in-depth treatment of the instrumentation, physical bases and applications of X-ray photoelectron spectroscopy (XPS) and static secondary ion mass spectrometry (SSIMS) with a specific focus on the subject of polymeric materials.

XPS and SSIMS are widely accepted as the two most powerful techniques for polymer surface chemical analysis, particularly in the context of industrial research and problem solving. Following a decade of rapid advances in instrumental capabilities and data interpretation, the field has reached a stage where it is possible to consolidate that information. In this book, the techniques of XPS and SSIMS are described and in each case the author explains what type of information may be obtained. The book also includes details of case studies emphasising the complementary and joint application of XPS and SSIMS in the investigation of polymer surface structure and its relationship to the properties of the material.

This book will be of value to academic and industrial researchers interested in polymer surfaces and surface analysis.

Surface analysis of polymers by XPS and static SIMS

Cambridge Solid State Science Series

EDITORS

Professor D.R. Clarke
Department of Materials,
University of California, Santa Barbara

Professor S. Suresh
Department of Materials Science and Engineering,
Massachusetts Institute of Technology

Professor I.M. Ward FRS
IRC in Polymer Science and Techonology,
University of Leeds

Surface analysis of polymers by XPS and static SIMS

D. Briggs
Siacon Consultants Ltd

CAMBRIDGE
UNIVERSITY PRESS

PUBLISHED BY THE PRESS SYNDICATE OF THE UNIVERSITY OF CAMBRIDGE
The Pitt Building, Trumpington Street, Cambridge CB2 1RP, United Kingdom

CAMBRIDGE UNIVERSITY PRESS
The Edinburgh Building, Cambridge CB2 2RU, United Kingdom
40 West 20th Street, New York, NY 10011–4211, USA
10 Stamford Road, Oakleigh, Melbourne 3166, Australia

First published 1998

Printed in the United Kingdom at the University Press, Cambridge

Typeset in 10¼ on 13½pt Monotype Times   [SE]

A catalogue record for this book is available from the British Library

Library of Congress cataloguing in publication data

Briggs, D. (David), 1948–
    Surface analysis of polymers by XPS and static SIMS / D. Briggs.
        p.   cm. – (Cambridge solid state science series)
    Includes bibliographical references.
    ISBN 0–521–35222–3 (hc)
        1. Polymers–Surfaces–Analysis.   2. X-ray spectroscopy.
    3. Secondary ion mass spectrometry.   I. Title.   II. Series.
    QD381.9.S97B75   1998
        620.1′92–dc21   97–26059 CIP

ISBN 0 521 35222 3 hardback

To Jill and Cherry

Contents

Preface

It is interesting to note that when this monograph was first contemplated, I had two relevant titles from the Cambridge Solid Science Series on my bookshelves: *Polymer Surfaces* by Cherry (now out of print) and *Modern Techniques of Surface Science* by Woodruff and Delchar. The former does not discuss surface analysis techniques whilst the latter does not mention application to polymers!

The behaviour of polymer surfaces is important in many technologies and understanding this behaviour requires surface characterisation with a high degree of chemical specificity, in terms of composition and structure, for species covering a wide range of molecular weight. The application of X-ray photoelectron spectroscopy (XPS or ESCA) and, later, static secondary ion mass spectrometry (SSIMS), in the early stages of the development of the techniques, to polymer surface analysis surely count as major successes. The requirements for polymer surface analysis have continued to be important drivers in the evolution of instrumental capabilities, because of the importance of this materials sector.

There is evidence that the polymer surface analysis field is now consolidating, following a period of major developments in instrumentation, spectroscopic performance and spectral interpretation. Spectral databases have been published, application oriented papers outweigh fundamental papers in the literature and instrument performance appears to have reached a plateau. The relative strengths of XPS and SSIMS are widely appreciated and there are many available reviews which cover, in more or less detail, the application of either technique, *separately*. It is clear, however, that most polymer surface studies benefit from the combined use of XPS and SSIMS, particularly now that time-of-flight (ToF)SIMS instruments are becoming more widely available. This seems an appropriate time, therefore, to attempt to treat the subject of polymer surface analysis using modern surface science techniques.

The book is organised as follows. Chapter 1 introduces polymer surfaces, describes why XPS and SSIMS have assumed their current importance within the set of potentially useful analytical techniques and gives their brief histories. Chapters 2 and 4 describe, respectively, the instrumentation and physical bases of the two techniques, together with important aspects of polymer-specific studies. Chapters 3 and 5, similarly, discuss the information obtained from polymer surfaces and how this is interpreted. Finally, Chapter 6 describes, in some detail, a series of case studies which illustrate the ways in which XPS and SSIMS are used to investigate polymer surface structure and behaviour *in the real world*, with emphasis on studies which combine the two techniques.

In order to keep the book reasonably concise, a basic level of chemistry (including polymer chemistry) and physics is assumed. This means that the material should at least be accessible to undergraduate students in the later stages of chemistry/materials science courses. The primary aim, however, is to treat the subject in a way which is useful to those already working in the polymer surface/analysis field and which will allow those wanting to enter the field at post-graduate level an up-to-date coverage of the subject.

I have been fortunate to have spent most of my research career to date in the subject area of this monograph. Much of the material used derives from this work and I wish to acknowledge the contribution of ICI and my former colleagues, especially Drs Graham Beamson, Ian Fletcher and Martin Hearn, as well of that of Prof. Buddy Ratner and his colleagues at the University of Washington. A significant fraction of the figures and tables are reproduced from the following publications of John Wiley & Sons (Chichester, UK): the journal *Surface and Interface Analysis* and the books *Practical Surface Analysis*, second edition, vols 1 & 2 (Eds. D. Briggs and M.P. Seah) and *High Resolution XPS of Organic Polymers* (G. Beamson and D. Briggs). I gratefully acknowledge their permission to reproduce this material.

I am indebted to Drs Martin Seah and Graham Leggett for each reading one of the draft chapters and making valuable comments and suggestions for improvement. Finally, my special thanks to Prof. Ian Ward to whom the emergence of this book owes so much. As a series editor he pursued the objective of a book on polymer surface analysis for many years. When I accepted his invitation, I little realised how long the project was going to last! His patient enquiries and encouragement kept me going and he also read, and made detailed comments on, the whole of the manuscript during its preparation.

Malvern   D. Briggs
August 1997

Abbreviations

AES	Auger electron spectroscopy
AFM	atomic force microscopy
ARXPS	angle resolved X-ray photoelectron spectroscopy
CAE	constant analyser energy
CCD	charge coupled detector
CHA	concentric hemispherical analyser
CMA	cylindrical mirror analyser
CRR	constant retard ratio
EELS	electron energy loss spectroscopy
ESCA	electron spectroscopy for chemical analysis
ESD	electron stimulated desorption
ESIE	electron stimulated ion emission
FRS	Rutherford forward recoil scattering spectroscopy
fwhm	full width at half maximum
GIXRD	grazing incidence X-ray diffraction
IRS	infrared spectroscopy
ISS	ion scattering spectroscopy
LEED	low energy electron diffraction
NEXAFS	near edge X-ray absorption fine structure spectroscopy
NR	neutron reflectometry
PSD	position sensitive detector
RBS	Rutherford back scattering spectroscopy
SE	spectroscopic ellipsometry
SEM	scanning electron microscope

SIMS	secondary ion mass spectrometry
SSIMS	static secondary ion mass spectrometry
STM	scanning tunnelling microscopy
TEM	transmission electron microscopy
ToF SIMS	time-of-flight secondary ion mass spectrometry
UHV	ultra-high vacuum
XPS	X-ray photoelectron spectroscopy
XRF	X-ray fluorescence
XRR	X-ray reflectometry

Chapter 1

Introduction

1.1   The importance of polymer surfaces

We are surrounded by polymeric materials. 'Plastics' have been replacing 'traditional' materials such as metals, wood, glass, paper, leather, etc. ever since the introduction of the first thermosetting resins – and the trend continues. Polymers are therefore pervasive: in the form of mouldings, sheets, fibres and films; in protective coatings (particularly paints), adhesives, sealants and printing inks; in composites with inorganic components as structural materials (e.g. glass fibre/polyester resin for boat hulls or carbon fibre/epoxy resin for aircraft sections). These aspects of polymer application are evident to all. Rather less evident generally is the rapidly increasing importance of polymers in biomedical applications. However, as early as 1981 the annual usage of, for instance, contact lenses, blood bags and catheters was approximately 2, 30 and 200 million items, respectively (Ratner, Yoon & Mates, 1987). Finally, there is the diverse variety of polymers with entirely novel properties which make possible recent or emerging devices (e.g. conducting polymers, optical fibre coatings, drug-release vehicles, liquid crystal displays).

Having established the importance of polymeric materials, *per se*, it is necessary to emphasise the role of the surface. In many cases surface properties are critical to the end-use or performance of the polymeric article. These include properties related to adhesion (e.g. wetability, printability, adhesive bonding, heat sealability, 'blocking', releasability), electrical properties (e.g. static chargeability, triboelectric behaviour, charge storage capacity), wear properties (e.g. friction, lubricity, wearability), optical properties (haze, gloss, stains etc.),

1

biological compatibility (a catch-all for a variety of responses to blood, tissue etc.), permeability, chemical reactivity and crazing. These properties are dependent upon the detailed physical and chemical structure of the polymer surface (the dimensions of this critical region are considered in Section 1.5).

1.2   Differences between polymer surface and bulk

Even for a pure linear homopolymer thermoplastic there are several potential differences between the surface and the interior (bulk or average). Polymer chains have end-groups which may be functionally quite different from the repeating unit (especially if they result from the incorporation of initiating and terminating species used during the polymerisation process). The chains are characterised by a molecular weight distribution and for certain polymers (e.g. polyamides) very low molecular weight oligomers may be present preferentially as cyclic, rather than linear, species. Many polymers are semicrystalline, possessing regions which are microcrystalline (usually consisting of at most 50% of the material, organised into spherulitic structures) and regions which are amorphous. End-groups tend to be excluded from crystallites; they may even confer surface activity. In principle, therefore, the homopolymer surface may differ from the bulk in end-group concentration, molecular weight distribution and, where relevant, amorphous:crystalline ratio. Deviations from complete linearity of the polymer chain arising from branching or cross-linking introduce additional complexity.

Copolymers have increased potential for surface:bulk differentiation, particularly in the case of block and graft copolymers. Here there is a tendency for like sections of the polymer structure to associate, leading to the development of domains observable by microscopy (often at only very high resolution in transmission electron microscopy, TEM) and the possibility that one component may dominate the surface region. Polymer blends (or alloys) represent this effect *in extremis*, since few polymers are actually compatible.

For thermoset polymers, with their extended three dimensional cross-linked network structure, the situation is somewhat simplified. Nevertheless, the cross-link density at the surface may differ from that of the bulk due to preferential segregation of either prepolymer or cross-linking agent (or catalyst if used), or to poor component mixing.

1.3   Effects of additives and contaminants

The plastic materials in everyday use are rarely 'pure' in the chemical sense. The base polymeric component is usually compounded, prior to processing, with various additives which are required either to aid the processing step itself or to impart various attributes to the final article which are not characteristic of the polymer. These additives range from small organic molecules to inorganic particles. Table 1.1 lists most of the categories of additives. Clearly they are very numerous. To give just one illustration of the complex formulation of a plastic article, Table 1.2 lists the typical contents of polypropylene mouldings used in car (automobile) bumpers and interior trim (e.g. facia). The additives in Table 1.1 are largely associated with thermoplastics. In polymer-based coatings other additives are used to promote flow, prevent sag, give surface levelling etc.

Table 1.1 *Common types of additives used in plastics*

Additive	Sub-type	Example
antimicrobial		copper-8-hydroxyquinolate
antioxidant	primary	2,6-di-tert-butyl-p-cresol
	secondary	tris (nonylphenyl) phosphite
antistat	external	cetyl trimethyl ammonium chloride
	internal	glyceryl monostearate
colourant	pigment, inorganic	titanium dioxide
	pigment, organic	phalocyanine derivative
	dye	anthraquinone derivative
coupling agent		aminopropyl trimethoxysilane
filler		calcium carbonate
flame retardent	inorganic	antimony oxide
	polymeric	poly(tribromostyrene)
	molecular	brominated diphenyloxide
foaming agent		1-1' azobisformamide
heat stabiliser	primary	di-n-octyl tin bis(iso-octylthioglycolate)
	secondary	butyl-9,10-epoxystearate
lubricant		ethylene bis-stearamide
mould release agent		zinc stearate
organic peroxide	initiator	di-sec-butyl peroxydicarbonate
	curing agent	lauroyl peroxide
plasticiser		di-octyl phthalate
smoke suppressant		aluminium trihydrate
UV stabiliser		2-hydroxy,4-alkoxybenzophenone derivative

In several cases, e.g. antistatic, lubricant and repellency agents, the additive is *intended* to migrate to the surface. However, in other cases, e.g. antioxidants and plasticisers, the additive is added to modify the bulk but may, under certain circumstances, surface segregate. In the case of film products, which are reeled, the two sides may be different and additive migration between them ('offsetting') may occur. Surface active agents may be inadvertently present in the polymer bulk either as residues from the polymerisation process (emulsion or suspension) or from their use as dispersants for particulate additives. Thus, there are numerous possibilities for the unintended presence of additives at the surface. Since surface behaviour can be markedly affected by them, these unintended molecules are generally regarded as 'surface contamination'. Surface contamination by external agencies also causes major problems. The processing versatility of polymers has, of course, resulted in large scale automatic and semiautomatic production of polymer-based articles. The machinery involved is associated with lubricating oils, greases, hydraulic fluids, vacuum pump oils etc., all of which may end up on the article surface. Transport of these molecules through the air either as vapours or as aerosol particles is a most important route to contamination. Contamination by direct contact is often unavoidable and even deliberate, e.g. in the application of sizes and finishes in fibre production or in the spraying of moulds with release agents.

1.4   Polymer surface pretreatment and modification

The fact that surface behaviour could seriously compromise the end-use of polymeric materials with otherwise excellent bulk properties became evident very

Table 1.2   *Composition of polypropylene mouldings found in automotive applications*

Component	Concentration (wt%)	Remarks
base polymer		usually a blend of homopolymers and copolymers (with ethylene) of variable molecular weights to give desired properties
mineral filler(s)	up to 25	usually talc, may be mixture
rubber modifier	up to 35	often a blend of ethylene copolymers
process, UV and heat stabilisers	up to 2	up to six different additives in the stabiliser package
pigment(s)	up to 1	several usually required to achieve desired colour

early in their development and application – low density poly(ethylene) film developed as a flexible, transparent, heat-sealable packaging material in the 1950s could not be printed with available inks. Low density polyethylene (LDPE) is hydrophobic (critical surface tension of wetting, $\gamma_c = 31\,\mathrm{mN\,m^{-1}}$) and therefore difficult to wet by ink solvents. It is now known that whilst wetting is a necessary condition for good ink adhesion, it is not sufficient. Also required are specific interactions of the acid–base type (of which H-bonding is the most commonly encountered). LDPE, being a purely hydrocarbon polymer, has no functionality. Empirical 'pretreatments' were developed which introduced surface functionality and overcame the problem – although even a general understanding of their efficiency only emerged, after many years of argument, with the application of the techniques discussed herein (see Section 6.4).

The general requirement to 'surface engineer' polymeric materials in order to overcome the frequently inadequate inherent surface properties has led to the development of many surface modification techniques. These are listed in Table 1.3. Understanding and monitoring these processes are important aspects of polymer technology (Brewis, 1982; Chan, 1993).

1.5 Depth scales associated with surface behaviour

Classical surface science is dominated by studies of metals and semi-conductors, usually in the form of single crystals. Structure-sensitive surface probes have provided overwhelming evidence that the transition from surface to bulk properties occurs within the first few atomic layers so that studies of

Table 1.3 *Treatments used to modify plastic surfaces, particularly to improve adhesion performance*

Surface treatment	Effect
solvent wipe	cleaning, roughening
corona discharge	oxidation
flame	oxidation
acid etching	oxidation, roughening
inert gas plasma	cross-linking
active gas plasma	functionalisation
plasma deposition	cross-linked thin polymer coating
surface grafting	attachment of different polymer chains
sodium naphthalenide/THF (for PTFE)	defluorination/oxidation

surface phenomena require techniques which are either (ideally) specific to a region ~1 nm below the surface or which can at least provide a high degree of contrast between this region and the bulk. It might be thought that in the case of polymers, the transition from surface to bulk properties would take place over length scales of the order of polymer chain sizes, i.e. up to several tens of nanometres, leading to less stringent sampling depth requirements. For some properties this is certainly true, but it does not necessarily follow that polymer surface phenomena are governed by the structure and composition within these same dimensions. In fact, there are very few definitive studies of polymer surface structure–property relationships. This is due to the difficulties associated with both the reproducible preparation of polymer surfaces with controlled structural and compositional features, and with their characterisation. One property which has been studied in detail, using self-assembled monolayer molecular films as models of polymer surfaces, is wetting. This work demonstrates that wetting behaviour is dominated by a layer of thickness ~0.5–1 nm (Allara, Atre & Parikh, 1993). As the techniques described in this book are applied to studies of wetting, adhesion, friction, biocompatibility etc., and particularly the effect of surface contamination on these behaviours, the general importance of this layer thickness becomes increasingly apparent.

1.6   Requirements for polymer surface analysis techniques

The ideal single technique would possess the following attributes: quantitative molecular speciation (including sensitivity to conformation etc.), sampling depth variability from 0.2 to 10 nm, lateral resolution of $<0.1\,\mu$m, *in situ* operation (e.g. in air, water ambients), insensitivity to surface roughness, real time analysis (fast measurements). It is, of course, taken for granted that the surface under investigation is unaffected by the measurement.

A great many ultra-high vacuum (UHV) surface science techniques have been developed during the last 30 years, but few satisfy the essential criterion from the above list, *viz* molecular sensitivity. On the other hand there exists a variety of solid state molecular spectroscopies, but few of these are surface sensitive. Table 1.4 describes those techniques which, in principle, are applicable to polymer surface characterisation in terms of some of the above attributes. None of these techniques remotely approaches ideality.

Bearing in mind the importance attached to probing surface chemistry with a high degree of discrimination, X-ray photoelectron spectroscopy (XPS) and static secondary ion mass spectrometry (SSIMS) stand out from the rest. Being vacuum techniques they lack *in situ* capability. Atomic force microscopy (AFM) probes topography and local material properties, with close to atomic resolution, but not chemical composition (other scanning probe microscopy techniques are being developed which will be able to provide some chemical/structural information). Infrared spectroscopy (IRS) is also, in principle, capable of *in situ* studies but is only capable of any degree of surface sensitivity in the internal reflection mode of operation, when the polymer surface is pressed into contact with the reflection element (e.g. a germanium crystal). For the surface analysis of real-world polymer systems, therefore, it is not surprising that XPS and SSIMS have become the dominant techniques. They are also highly complementary and are increasingly used together in both fundamental and applied investigations. The rest of this book is concerned with these methods.

1.7 Brief history of XPS

XPS dates back about a century to the early days of atomic physics, following the discovery of the photoelectric effect by Hertz in 1887. Before the First World War several groups analysed the energies of electrons emitted from metals bombarded by hard X-rays, the most prominent of which was Rutherford's group in Manchester. This group had extensive experience of the measurement of X-ray spectra from radioactive materials using magnetic analysis and was, furthermore, at the forefront of X-ray spectroscopy. By 1914 Rutherford came close to stating the basic XPS equation:

$$E_K = h\nu - E_B - \phi$$

where E_K is the photoelectron kinetic energy, $h\nu$ the exciting photon energy, E_B the electron binding energy in the solid and ϕ the work function (Rutherford omitted this final term).

Robinson, alone, from this group continued research after the war and Maurice de Broglie began work in France, still using similar techniques with photographic detection. Understanding of core level spectroscopy developed rapidly but the true nature of the 'anomalous features' now known as Auger series only became clear in 1925 from Pierre Auger's entirely unrelated cloud chamber experiments. Gradually, interest in XPS faded as X-ray spectroscopy developed into the superior technique for atomic structure investigation.

After the Second World War Steinhardt and Serfass at Lehigh University revived XPS with the aim of performing surface chemical analysis. Although their instruments did not lead to improved performance, they did report surface effects on spectra in 1951 (despite using very high energy X-rays). Up to this point the spectra consisted of a series of bands with a reasoanbly well-defined

Table 1.4 Applicability[a] of technique to analyse a ≤5 nm surface region for different sample types (adapted from Allara et al., 1993)

Technique[b]	Bulk polymer			Smooth, thin (<50 nm) polymer film on a planar substrate	Molecular monolayer surface on a planar substrate
	Rough, non-planar surface; bulk and surface very similar	Highly smooth, planar surface; bulk and surface very similar	Highly smooth, planar surface; bulk and surface very different		
XPS	+	+	+	+	+
SSIMS	+	+	+	+	+
EELS	+	+	+	+	+
NEXAFS	+	+	+	+	+
AFM[c,d]	+	+	+	+	+
ISS		+	+	+	+
RBS/FRS			+	+	+
IRS[d]			+	+	+
SE[d]			+	+	+
Raman[d]			+	+	+
NR[d]				+	+
XRR[d]				+	+

GIXRD[d]	+	+
He scattering		+
LEED		+
STM[d]		+

[a]Qualitative assessment of general applicability. In some cases extraordinary circumstances may allow an analysis where no + is given.

[b]Definition of technique acronyms:

Spectroscopies: XPS, X-ray photoelectron; SSIMS, static secondary ion mass; EELS, electron energy loss; NEXAFS, near edge X-ray adsorption fine structure; ISS, ion scattering; RBS/FRS, Rutherford back scattering and forward recoil scattering; IRS, infrared; SE, ellipsometry.

AFM, atomic force microscopy; NR, neutron reflectometry; XRR, X-ray reflectometry; GIXRD, grazing incidence X-ray diffraction; LEED, low energy electron diffraction; STM scanning tunnelling microscopy.

[c]And related techniques.

[d]Non-vacuum techniques.

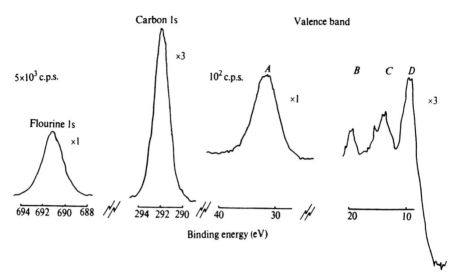

Figure 1.1   Core and valence levels of PTFE, the first XPS spectrum from a polymer surface (Clark & Kilcast, 1971).

edge followed by a tail. However, following major advances in β-ray spectroscopy at the University of Uppsala during the 1940s, Kai Siegbahn constructed an XPS instrument capable of measuring electron kinetic energy with a resolving power of 10^5. In 1954 discrete peaks on the high kinetic energy side of the band 'edges' were fully resolved, corresponding to core electrons emitted with zero energy loss. This allowed E_K (and hence E_B) to be measured accurately for the first time and soon after core level 'shifts' resulting from chemical state variation were observed. Siegbahn coined the acronym ESCA (electron spectroscopy for chemical analysis), which is still in widespread usage, and made this the title of a book (Siegbahn *et al.*, 1967). This described the development of XPS in terms of instrumental developments, application to atomic, molecular and solid state structure, and theoretical frameworks for interpretation, by the Uppsala group, over more than a decade. It had an immediate and widespread impact, although surface analysis hardly featured.

Commercial instruments started to appear around 1969–70 with the construction of the first UHV system in 1972. At this time the real surface sensitivity of XPS was appreciated. There are no inherent problems in studying polymers and the first spectrum to be reported was that of poly(tetrafluoroethylene) (PTFE) (Clark & Kilcast, 1971) (Fig. 1.1). Clark and coworkers at the University of Durham then went on to study systematically the information levels in the XPS spectra of polymers during the 1970s. Step changes in instrumental performance had to wait until the late 1980s with the introduction of instruments giving high sensitivity with high energy resolution (Gelius *et al.*, 1990) and direct imaging at $<10\,\mu m$ resolution (Coxon *et al.*, 1990).

1.8   Brief history of SIMS

The origins of SIMS also date back to early days of atomic physics. During his famous experiments with discharged tubes, J.J. Thomson (in 1910) studied the effects of the 'positive rays' (ions) on a metal plate. He deduced that the secondary particles emitted in all directions were mainly neutral species, but that a small fraction were positively charged. The early studies of positive rays and of their deflection by electric and magnetic fields led to the birth of mass spectrometry. In 1913 J.J. Thompson first applied this technique to study elemental isotopes. Mass spectrometry advanced rapidly on several fronts and the first negative ion spectra were reported by Woodcock in 1931 (for sodium and calcium fluorides and with approximately unit mass resolution). SIMS instruments built for analytical purposes were developed from 1949 by Herzog and Viehbock and from 1950 by Honig and others at RCA Laboratories. In fact, Herzog and coworkers built the first commercial SIMS instrument under a 1960s NASA contract for the analysis of the spatial and isotopic distribution of all elements in samples brought back from the moon. This led to an appreciation of the power of SIMS for the analysis of such materials and its advantages for microanalysis over electron probe methods. The rise of the semiconductor industry with its need for in-depth, high sensitivity microanalysis ensured rapid development of high spatial resolution instruments based on both microprobe and microscope approaches. In all of these applications material is consumed during analysis either by necessity (high sensitivity to trace elements) or by design (depth profiling) and surface analysis is not the aim. This field has come to be known as dynamic SIMS.

True surface analysis by SIMS dates back to 1969 and the very different approach of Benninghoven at the University of Münster. He applied the technique to study surfaces in UHV by deliberately using low primary ion currents covering large areas. The early quadrupole mass spectrometers had sufficient sensitivity to allow spectra to be acquired well within the sputtering time for the first monolayer (and to be representative of virgin surface); Benninghoven therefore distinguished this approach from the earlier work by the term static SIMS (SSIMS) (Benninghoven, 1969). The work of the Münster group on adsorbed layers of delicate organic molecules (on metallic substrates) indicated that the analytical power of organic mass spectrometry might be brought to bear on polymer surfaces (Gardella & Hercules, 1980).

Unlike XPS, however, the application of SIMS to polymeric materials was not straightforward. Ion beam damage rates are greater than would be expected from monolayer sputtering rates and insulator charging effects are serious. It was not until 1982 that work at ICI Plastics Division with a specially designed instrument (incorporating a quadrupole mass spectrometer with high transmission and mass range) resulted in the first spectra to be obtained under well-defined

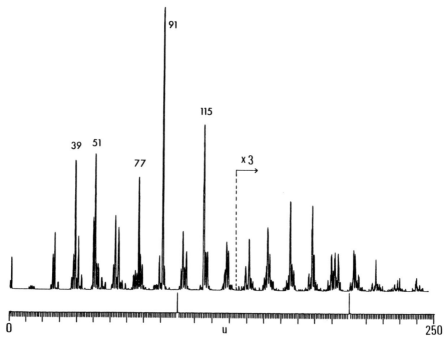

Figure 1.2   Positive ion spectrum of poly(styrene), the first truly static SIMS spectrum from a polymer surface (Briggs & Wootton, 1982).

truly 'static' conditions (Briggs & Wootton, 1982; Briggs, 1982) (Fig. 1.2). Briggs and coworkers systematically developed the field during the 1980s, having demonstrated molecular imaging by SIMS at a fairly early stage (Briggs, 1983) (Fig. 1.3). The major instrumental developments have been in the mass spectrometry, with the introduction of time-of-flight designs, again by Benninghoven's group, resulting in orders-of-magnitude increases in sensitivity, mass resolution and mass range compared with the early quadrupole instruments.

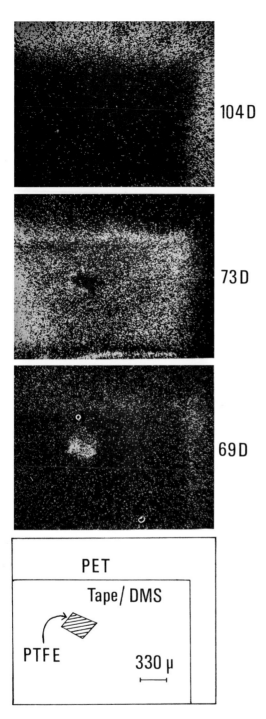

104 D

73 D

69 D

PET

Tape / DMS

PTFE

330 μ

Figure 1.3 One of the earliest examples of SIMS imaging from a multicomponent polymer surface. The species detected in the images are: m/z 104 ($C_6H_4CO^+$, unique to poly(ethylene terephthalate), PET); m/z 73 (($CH_3)_3Si^+$, unique to dimethylsilicone, DMS) and m/z 69 (CF_3^+ from PTFE but also $C_5H_9^+$ from PET and adhesive tape). As shown in the diagram the sample consisted a small piece of PTFE mounted on a larger piece of double-sided adhesive tape (with a silicone release agent) attached to the surface of PET film (Briggs & Hearn, 1985).

Chapter 2

XPS

2.1 Instrumentation

XPS involves the irradiation of a sample with soft X-rays and the energy analysis of photoemitted electrons which are generated close to the sample surface. An X-ray photoelectron spectrometer (or XPS instrument) therefore consists of: a vacuum vessel with its associated pumping system and sample introduction/manipulation systems; an X-ray source; an electron energy analyser and its associated input (or transfer) electron–optical lens; an electron detection system and a dedicated 'datasystem' based on a PC or workstation which both controls the spectrometer operation and provides the means to process the acquired data. These components are now considered in turn.

2.1.1 Vacuum system

A comprehensive electron spectrometer vacuum system of typical early 1980s design is illustrated in Fig. 2.1. The spherical analysis chamber allows several probes and analysers to be mounted so that several surface spectroscopies can be applied to the same sample. The construction is mostly of stainless steel with joints made by compressing knife-edge flanges against copper gaskets. In XPS it is important to screen electron trajectories from the earth's magnetic field and this is achieved either by fabricating the analyser chamber and electron analyser/lens vessels out of a material with a high magnetic permeability, usually mu-metal, or by cladding the steel vessel with mu-steel sheeting (internally or externally).

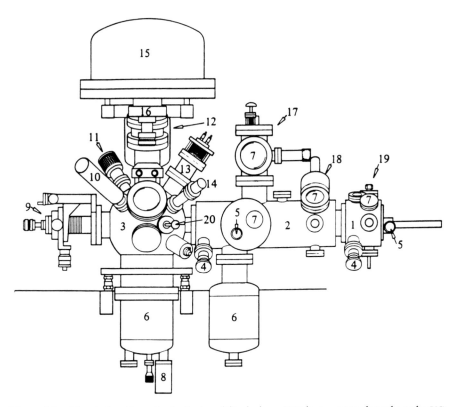

Figure 2.1 Front elevation of a typical multitechnique XPS instrument, based on the VG ESCALAB Mk II design: 1 fast entry specimen insertion lock (stainless steel); 2 UHV specimen preparation chamber (stainless steel); 3 UHV experimenta/analysis vessel (mumetal); 4 viton-sealed gate valve; 5 rotary drive to specimen transfer mechanism; 6 titanium sublimation pump vessel (stainless steel); 7 viewport; 8 autocarousel motor drive; 9 high precision (*X*, *Y*, *Z* translation and θ tilt) specimen manipulator; 10 twin anode X-ray source (Al/Mg); 11 UV discharge source (UPS); 12 monochormated X-ray source (Al/Ag) 13 2000 Å electron source (AES, SEM, SAM); 14 scanning ion source; 15 electron energy analyser vessel (mu-metal); 16 detector (single or multichannel); 17 specimen fracture stage; 18 static, broad beam ion source; 19 high pressure gas reaction/catalysis cell; 20 'Wobble stick' specimen transfer fork (After Christie, 1989).

In order for electrons to reach the analyser without being scattered by residual gas molecules, the achievable vacuum only needs to be about 10^{-5} Torr $(1.33 \times 10^{-3}\,\text{Pa})$, i.e. in the high vacuum regime. However, most vacuum systems routinely operative in the UHV regime ($<10^{-9}$ Torr) in order to reduce the contamination of sample surfaces by absorbed residual gas molecules. Assuming these molecules have a sticking probability of unity, then at 10^{-5} Torr a clean surface would be covered by a monolayer in about 0.2 s, whereas this process would take about 30 min at 10^{-9} Torr. Since sticking coefficients for polymer surfaces are usually much less than unity this an acceptable operating regime.

Several types of pumping system are available: liquid nitrogen trapped, water-cooled oil-diffusion pumps; turbomolecular pumps; sputter-ion pumps; titanium sublimation pumps; and cryopumps. The first three types are most frequently used for the main pumping system, whilst the others are most usually used as auxiliary pumps. All have advantages and disadvantages (Riviere, 1990a) depending on system design and use. Systems pumped in this way and baked-out overnight (100–160°C) will achieve a base pressure of 10^{-10} Torr or less. It is important to bake regularly systems used routinely for polymer studies since the (usually) slow loss of volatile components ensures that operating pressures are significantly above base pressure.

2.1.1   Sample handling

The system illustrated in Fig. 2.1 involves three vacuum chambers. The small insertion lock is usually pumped only by a rotary oil-pump (base pressure 10^{-3}–10^{-4} Torr). The sample is placed on a 'railway' transfer device with the insertion lock at atmospheric pressure. Pumpdown takes only a few minutes and, on opening a gate valve between this chamber and the second 'preparation' chamber, a rotary drive moves the sample into it. The second chamber is continuously pumped with a UHV base pressure. The sample is transferred using a 'wobble stick' onto a second railway which connects similarly with the analysis chamber via another gate valve. Since the second chamber acts as a vacuum buffer the analysis chamber can be maintained under UHV conditions despite frequent sample introductions. For the same reason the second chamber is frequently used for sample 'preparation' or conditioning (hence its popular name) such as heating, cleaning by ion bombardment or scraping, fracture or peeling for interface studies, plasma treatment etc. However, this arrangement owes much to a design philosophy driven by surface physics requirements (ultra-clean single crystal studies, gas adsorption experiments and the like). For more routine analytical operation involving samples from the 'real world' a two-chamber system is more appropriate and the vacuum conditions in the analysis chamber are slightly compromised. As noted above, the routine study of polymers has this effect anyway. The first chamber can be enlarged to accommodate sample 'preparation' facilities and pumped by a small turbomolecular pump to decrease pumpdown time and lower base pressure. Such a system has the distinct advantage that sample introduction can be based on a rod pushed through the single gate valve, and this can be automated.

Once in the analysis chamber the sample is invariably transferred onto a precision manipulator with several degrees of freedom: at least X, Y, Z and tilt (with respect to the angle between sample surface and electron optics entrance aperture). Sometimes azimuthal rotation (in the sample surface plane) is added. The manipulator may accommodate a single sample stub or a plate holding many samples. Sample positioning may be manual or auto-mated (following setting of coordinates via the datasystem). Older designs often incorporate a carousel capable of holding several samples in fixed posi-tions around the perimeter of a disc which is rotated automatically to bring each sample into the analysis position in turn (also shown in Fig. 2.1). In very early instruments the sample is usually positioned on the end of an insertion rod pushed into the analysis chamber through an entry lock of varying sophistication.

One facility which is particularly useful for polymer studies is sample cooling during analysis. This is frequently difficult to achieve unless provision was included in the original design specification of the instrument; conse-quently it is often overlooked. The benefits of sample cooling are discussed in Section 3.3.

2.1.3 X-ray sources

X-rays are generated by bombarding a target (anode) with high energy elec-trons from a heated filament. The emission consists of characteristic lines, resulting from electronic transitions into the core-holes created by electron impact, superimposed on a continuous spectrum (Bremsstrahlung radiation). Thus for an aluminium target bombarded with 15 keV electrons the character-istic line is the unresolved $K\alpha_{1,2}$ doublet resulting from $2p_{3/2} \rightarrow 1s$ and $2p_{1/2} \rightarrow 1s$ transitions and this is about 100 times as intense as the Bremsstrahlung back-ground at that energy (1486.6 eV), as shown in Fig. 2.2. There are also weaker characteristic lines (or satellites) resulting from similar transitions within the multiply ionised atom (e.g. $K\alpha_3$, $K\alpha_4$) and from valence band $\rightarrow 1s$ transitions ($K\beta$). The relative intensities and positions of these satellites can be visualised from Fig. 2.3. Only two anode materials are commonly encountered in XPS. One is aluminium, the other is magnesium (characteristic line Mg $K\alpha_{1,2}$ at 1253.6 eV). These are the only materials which satisfy a number of criteria for X-ray sources:

(1) narrow linewidth to minimise contribution to photoelectron linewidth (Mg $K\alpha = 0.70$ eV, Al $K\alpha = 0.85$ eV fwhm);

(2) sufficient energy to excite photoemission from at least one core level for all elements;

(3) ease of fabrication of anode;

(4) high thermal conductivity for efficient heat dissipation at high power.

For efficient X-ray production, the electron energy needs to be about ten times the threshold energy. Thus for Al $K\alpha$ (about 1.5 keV) the anode potential is usually at 15 kV (the filament is at ground potential). Maximum power is

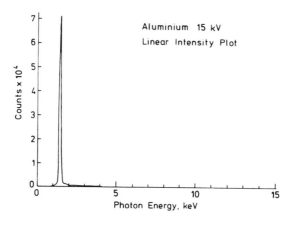

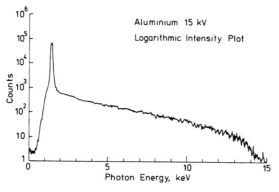

Figure 2.2   X-ray emission spectrum from an aluminium target bombarded with 15 keV electrons, showing the characteristic $K\alpha$ line superimposed on the Bremsstrahlung background (the detector used caused considerable broadening of the $K\alpha$ line) (Riviere, 1990a).

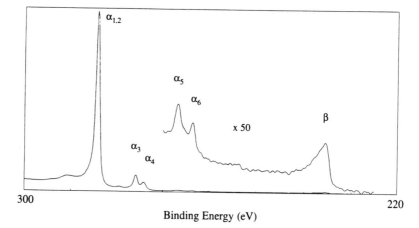

Figure 2.3   Magnesium X-ray satellites observed in the C 1s spectrum of graphite (Moulder *et al.*, 1992).

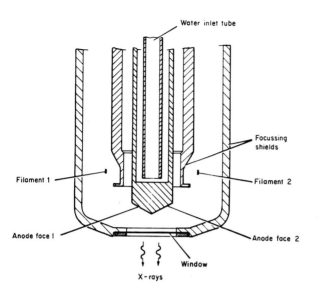

Water inlet tube

Focussing shields

Filament 1

Filament 2

Anode face 1

Anode face 2

Window

X-rays

Figure 2.4 Double-anode X-ray source (Barrie, 1977).

dictated by the target surface temperature (and hence potential for evaporation) and efficiency of water cooling. For static, single anode sources the operating limits are about 0.6 kW and 1 kW for magnesium and aluminium respectively. The highly popular dual (magnesium/aluminium) sources (Fig. 2.4) are typically restricted to about two-thirds these limits. The anode material forms a coating on a copper substrate and aged targets can give rise to O $K\alpha$ emission from the oxidised surface or to Cu $L\alpha$ from the copper base. The relative positions of these 'ghost' lines are included with the satellite lines in Table 2.1

Table 2.1 *X-ray satellites of Mg and Al $K\alpha_{1,2}$ and relative positions of the O and Cu $K\alpha$ ghosts on the kinetic energy scale*

X-ray line	Separation from $K\alpha_{1,2}$(eV) and relative intensity ($K\alpha_{1,2}=100$)	
	Magnesium	Aluminium
$K\alpha'$	4.5(1.0)	5.6(1.0)
$K\alpha_3$	8.4(9.2)	9.6(7.8)
$K\alpha_4$	10.0(5.1)	11.5(3.3)
$K\alpha_5$	17.3(0.8)	19.8(0.4)
$K\alpha_6$	20.5(0.5)	23.4(0.3)
$K\beta$	48.0(2.0)	70.0(2.0)
O $K\alpha$	−728.7	−961.7
Cu $K\alpha$	−323.9	−556.9

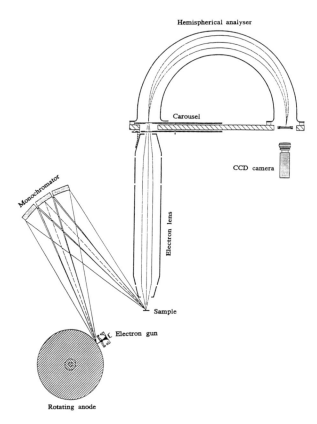

Figure 2.5
Monochromation of Al
Kα using a quartz crystals.
(Schematic of the Scienta
ESCA300 instrument,
Scienta AB, Sweden.)

2.1.4   Monochromated X-ray sources

It is possible to monochromate Al *Kα* radiation by means of a quartz crystal
monochromator; this results in the elimination of the Bremsstrahlung, satellite
and ghost radiation and allows the characteristic X-ray linewidth to be reduced
to below 0.3 eV. Al *Kα* radiation generated in the normal way is directed into a
quartz crystal slice oriented such that back-diffraction from the (10$\bar{1}$0) planes
occurs. From the Bragg relation:

$$n\lambda = 2d \sin \theta \qquad\qquad (2.1)$$

and for Al *Kα*, (λ=8.3 Å), quartz (10$\bar{1}$0) (2d=8.5 Å) a first order diffraction
(n=1) occurs at θ=78.5°. This allows for a convenient geometrical arrangement
since the angle between the incident and diffracted rays is 23°. Source, crystal
and sample are placed on the circumference of a Rowland circle, as shown in Fig.
2.5, such that an image of the X-ray source is focused onto the sample. Efficient
focusing demands that the crystal is bent and/or ground (Barrie, 1977; Chaney,
1987). The natural resolution of the quartz crystal is 0.16 eV but the different X-

ray energies in the original source are dispersed along the Rowland circle: e.g. the dispersion for a 0.5m diameter is 1.6mmeV^{-1}. Thus the final X-ray linewidth will depend on the dimension of the monochromated X-ray spot, on the sample, in the dispersion plane. Since only about 10% of the incident Al $K\alpha$ radiation is reflected onto the sample by the monochromator crystal(s), there is potentially a considerable loss of intensity relative to a conventional source. However, all the X-rays can be focused into a spot which matches the area of the sample viewed by the analyser optics and by the use of multiple crystals a large solid angle can be achieved. Use of a rotating anode source, as in Fig. 2.5 (Gelius et al., 1990), overcomes the power dissipation restrictions, noted above, for a static anode source. Overall, this allows the useful X-ray flux to be *increased* by an order of magnitude. Moreover, the removal of the Bremsstrahlung radiation reduces the background in the spectrum and significantly improves the signal:noise.

2.1.5 Electron energy analysers

The electron energy analyser measures the energy distribution of electrons emitted from the sample; the photoelectron spectrum is a plot of intensity versus kinetic energy. Only two types of analyser are in common usage, namely the cyclindrical mirror analyser (CMA) and the concentric hemispherical analyser (CHA). The latter, also referred to as the spherical sector analyser, is now universally employed in high performance XPS instruments. CMAs are particularly suited to Auger electron spectroscopy (AES), therefore in some combined XPS/AES instruments the photoelectron spectrum is produced by a CMA. Details of the design and performance of CMAs can be found elsewhere (Seah, 1989; Riviere 1990a). Here we concentrate on the CHA.

It is useful, at this stage, to define some terms relating to analyser performance. *Absolute* energy resolution, measured for a chosen peak in the spectrum, is defined either by the full width at half-maximum (fwhm, $\Delta E_{1/2}$ or simply ΔE) or as the base width (ΔE_b). The latter appears in many electron–optical equations, but the former is that usually quoted. Generally $\Delta E < 0.5\Delta E_b$. *Relative* energy resolution, R, is defined for a peak at a particular kinetic energy, E_0, by $R = \Delta E/E_0$ (or as a percentage, $100R$). The *resolving power* is $1/R$.

The essential features of the CHA are illustrated in Fig. 2.6. The two concentric hemispheres, radii R_1 and R_2, have a mean equipotential surface between them of radius R_0. Ideally $R_0 = (R_1 + R_2)/2$. Potentials $-V_1$ and $-V_2$ are applied to the hemispheres as shown with $V_2 > V_1$. If electrons of energy E_0 are injected at S tangentially to the equipotential surface they will be brought to a focus at F, irrespective of the plane of their circular trajectory, according to:

$$e\Delta V = e(V_2 - V_1) = E_0(R_2/R_1 - R_1/R_2) \tag{2.2}$$

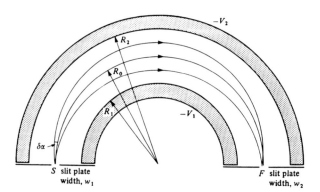

Figure 2.6 Cross-sectional view of a hemispherical electron energy analyser (Seah, 1989).

For electrons injected at an angle α to the correct tangential direction and with an energy $E > E_0$ under the same deflection potential (ΔV) conditions, the shift away from F along the radius R_0 (ΔR) is given by:

$$\Delta R = 2R_0[(E - E_0)/E_0 - (\delta\alpha)^2] \tag{2.3}$$

from which it is possible to derive an expression for the base resolution:

$$\frac{E_b}{E_0} = \frac{W_1 + W_2}{2R_0} + (\delta\alpha)^2 \tag{2.4}$$

where W_1 and W_2 are the entrance and exit slit widths. These are usually made equal ($W_1 = W_2 = W$). Clearly both slit width and entrance angle (defined as the half-angle, α radians) affect both the energy resolution (from (2.4)) *and* the sensitivity, whilst the relative values of the slit and angular terms affect the peak shape. The common compromise is to design the analyser such that $\alpha^2 < W/2R_0$ and this leads to an approximate expression for the relative resolution:

$$\frac{\Delta E}{E_0} \sim \frac{W}{2R_0} \tag{2.5}$$

Since the analyser dimensions are fixed it is clear from equation (2.2) that by ramping the deflection voltage (ΔV) electrons of progressively higher energy will be focused at F. Measuring electron intensity as the energy range is scanned produces the photoelectron spectrum.

In practice the electron–optical arrangement is rather more complicated. Firstly, it is not convenient to have the source of photoelectrons at S. An input (or transfer) lens is used primarily to transport electrons from the sample to the analyser in such a way that an image of the analysed area is projected onto the analyser entrance slit. This allows the sample to be conveniently placed away from the analyser; with an appropriately designed lens the lens–sample distance can be at least 50mm to provide plenty of working space around the

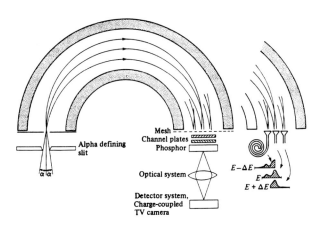

Figure 2.7 Two possible multichannel detector systems for spherical sector analysers: multiple channeltrons (right) and position-sensitive detector (left) (Seah, 1989).

sample/manipulator. Lenses of greater sophistication can provide small area analysis and imaging capability, as described in Section 2.1.7.

Secondly, the photoelectrons emitted by the sample are not analysed at their original energy, but are retarded before they enter the analyser. Often this is achieved by applying potentials to two closely spaced meshes, but it can also be performed by the input lens. This is necessary to achieve the required energy resolution with an analyser of reasonable size. If the analyser contribution to the peak width is to be small, then in the case of monochromated Al$K\alpha$ (source fwhm~0.3eV) the absolute resolution, ΔE, needs to be significantly less over the whole energy range. To achieve ΔE=0.15eV at 1500eV, the relative resolution needs to be 10^{-4}. From equation (2.5) if W=1mm, then to meet this performance criterion R_0 would have to be 5m; clearly this is impracticable. Electrons can be retarded without altering their absolute energy spread, therefore retardation allows an analyser of poorer resolving power to achieve the same ΔE. Retardation can be performed in two ways. Either all electrons are retarded to the same energy (so-called constant analyser energy, CAE, mode) or they are retarded to a certain fraction of their original energy (constant retard ratio, CRR, mode). The latter is the normal mode for AES, but the former is now near-universal for XPS. The analyser 'pass energy' in the CAE mode is selectable between, say, 5 and 200eV; energy resolution degrades as the pass energy increases whereas sensitivity increases. Returning to the above example: at a pass energy of 15eV, ΔE=0.15eV is obtained at a relative resolution of 10^{-2}, which is well within manufacturing capabilities.

2.1.6 Detectors

The simplest system is a single channel electron multiplier placed at F in Fig. 2.6. This is a coiled tube of semiconducting glass connected to a cone of about 1cm diameter at the open end (the shape is shown in Fig. 2.7), giving a gain of ca 10^8.

This is operated in ac mode (pulse counting) which provides single electron counting. Originally the amplified and discriminated pulses were fed to a rate-meter, with intensity measured in counts s^{-1}. Today, of course, the pulses are accumulated by the digital datasystem and intensity is generally displayed as counts channel^{-1}. Channeltrons run into saturation in the region of 10^5–10^6 counts s^{-1} and have a dark count rate of only 1–2 counts min^{-1}. The detector efficiency can depend markedly on electron energy.

The spherical sector analyser is, however, ideally suited to the use of multi-detector schemes. The transmitted electrons are dispersed in energy across the analyser exit plane by $2R_0/E_0$ mm eV^{-1} (E_0 is the pass energy). In principle, there-fore, replacement of the single exit slit and detector by a series of detectors and equivalent slits, arranged along the dispersion direction, will lead to a multiplex advantage through parallel acquisition. Fig. 2.7 shows two approaches for achieving this advantage. One is simply to increase the number of channeltrons (e.g. up to six for typical analyser dimensions). The other is to use a position-sensitive detector (PSD).

Typically, the exit slit is replaced by a pair of microchannel plates which consist of an array of channel multipliers a few tens of microns in diameter fused into a thin disc. Exiting electrons are detected as a function of position in a variety of ways (Hicks *et al.*, 1980) of which one, conversion to light pulses followed by their detection using a charge couple detector (CCD) camera, is shown in Fig. 2.7. The outputs from the multiple detector channels are sorted by the datasystem com-puter so that they are added into the appropriate spectral (energy) channels. The actual advantage gained by multidetection using a PSD may be less than expected on the basis of increased active detector area because of lower saturation count rates, higher background count rates and additional spectral noise.

2.1.7   Imaging XPS

Selected area analysis can be achieved by the use of a fine focus crystal mono-chromator as described in Section 2.1.4. With a highly focused electron source the X-ray spot size on the sample may be as low as 10 μm (but only, of course, for Al$K\alpha$ radiation). Alternatively, using a conventional flood X-ray source, a small area for analysis can be selected by electron–optical aperturing, i.e. pho-toelectrons are imaged through the analyser input lens onto an aperture which defines the area on the sample from which electrons enter the electron energy analyser (Christie, 1989). With the development of magnetic lenses which dramatically improve photoelectron collection efficiencies, the defined area can be ~10 μm in diameter (Drummond, 1992).

Both approaches to selected area analysis have been developed to produce an imaging capability. In the case of the former, a focused electron beam is scanned over the anode to generate, via the monochromator, a scanning X-ray probe

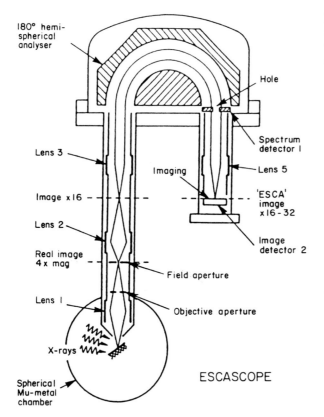

Figure 2.8 Parallel imaging XPS instrument, the VG ESCASCOPE (Coxon *et al.*, 1990).

(Larson & Palmberg, 1995). In the case of the latter, deflector plates are positioned in front of the transfer lens so that the defined area is moved and an image built up sequentially (Seah and Smith, 1988). In both cases, chemically specific images (intensity as a function of position) are acquired by tuning the analyser to pass electrons from a peak in the photoelectron spectrum which represents the element of interest.

A parallel imaging system is shown in Fig. 2.8 (Coxon *et al.*, 1990). Conventional optics are used to generate a magnified image of the surface. Lens 3 projects the image to infinity so that electrons of all energies that pass through the spectrometer are emitted parallel to the axis to be imaged by lens 5. In this way the image quality is not degraded by the dispersing effect of the CHA. Six channeltrons, arranged as in Fig. 2.7 but about a central hole, collect the energy spectrum (detector 1). For any chosen photoelectron energy, the analyser can be set up so that these electrons pass through the central hole. Lens 5 reverses the image transformation of lens 3 and a two dimensional image is collected on a PSD (detector 2). Peak and local background images are collected, stored in a framestore and then subtracted to produce an image of the distribution of the chosen species over the sample area. The inherent spatial resolution is $\sim 2\,\mu$m.

PHOTOGRAPH OF
AU/SI SAMPLE

SPATIALLY RESOLVED ESCA SPECTRA

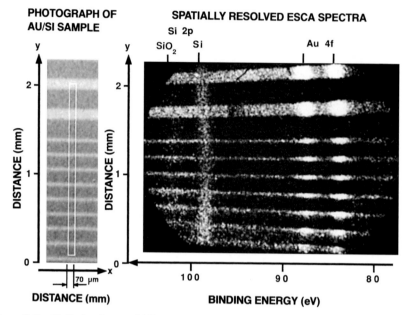

DISTANCE (mm)

BINDING ENERGY (eV)

Figure 2.9   *E–X* plot from gold lines on a silicon wafer. Left: diagram of sample with 32 μm and 13 μm gold lines with analysed area (0.07×2.1 mm shown vertically). Right: photograph of the multichannel detector output to a video monitor with the analyser set so that the binding energy range ~80–150 eV is detected. Thus each gold line gives rise to a spectrum dominated by the Au4f doublet and its high binding energy tail whereas the regions between the lines show the silicon and weaker oxide layer Si 2p lines (Beamson *et al.*, 1990).

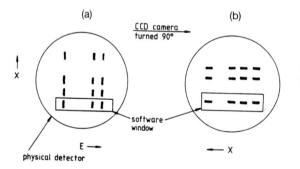

Figure 2.10
Microanalysis modes: (a) Energy scan of selected area. By stepping the software window position a microanalytical profile across a sample feature is built up. (b) Line scan at a selected energy (Beamson *et al.*, 1990).

Without lens 3 in Fig. 2.8 it is more traditional to image the sample at the entrance slit of the CHA. The hemispherical analyser itself can then only produce a one dimensional image with the other dimension's data convolved with the energy dispersed data. For a region of the sample imaged onto a narrow analyser entrance slit by a magnifying input lens, the image on the PSD is an '*E–X*' plot. Each pixel along the slit gives rise to an energy spectrum in the orthogonal dispersion plane, the pixel resolution being determined by the slit dimensions, lens magnification and spectrometer resolution. This is illustrated in Fig. 2.9. Fig. 2.10

shows how the use of a software window allows microanalysis or linescans at a chosen peak energy to be obtained (Beamson *et al.*, 1990). Finally, by operating in microanalysis mode (i.e. full spectrum from one pixel) and physically stepping the sample a complete image set covering, if necessary, the whole photoelectron spectrum can be acquired. The spatial resolution can be $<10\,\mu m$ (Gelius *et al.*, 1990). All four approaches to imaging just described are used commercially.

2.1.8 Datasystems

For many years now all instruments sold have been controlled by computer, either of the PC or workstation category, and most functioning instruments will have been upgraded if originally only manually controlled. Many of the advanced design features discussed in the preceding sections concerned with multidetection and imaging are only possible because of the computer. Digital voltage setting also means that all the optimum spectrometer conditions associated with a particular experiment can be recalled virtually instantaneously from memory to be reexecuted at will. Computer control of all the sample manipulator degrees of freedom allows complex sequences of analyses (e.g. multisampling of a large area to ascertain hetereogeneity, total analysis of many individual samples, analysis at many take-off angles, ion-etching/analysis cycles for depth profiling) to be preprogrammed and allowed to run unattended.

Following acquisition the computer then performs all the tasks of data reduction, processing, analysis and communicating with final output devices. Processing and analysis will feature in the ensuing sections; they include spectral smoothing, background subtraction, differentiation, peak area measurement and quantification, deconvolution, curve fitting and factor analysis. These are discussed in detail by Sherwood (1990).

2.2 Physical basis

2.2.1 The form of XPS spectrum

The soft X-rays used in XPS penetrate many microns into materials. X-ray absorption by an atom in the solid leads to the ejection of an electron (photoionisation) either from one of the tightly bound core levels or from the more weakly bound valence levels (or molecular orbitals). Some fraction of these electrons emerges from the surface into the vacuum. Overall, this is known as the photoelectric effect. The photoelectron emission is energy analysed as described above to produce a spectrum of electron intensity as a function of energy, as shown in Figs. 2.11 and 2.12.

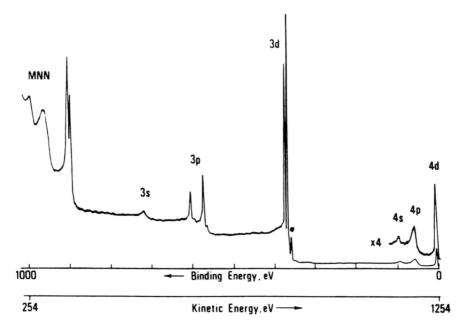

Figure 2.11 XPS spectrum of clean silver excited by $MgK\alpha$ (essentially $MgK\alpha_{1,2}$). * is the Ag 3d 'satellite' excited by $MgK\alpha_{3,4}$.

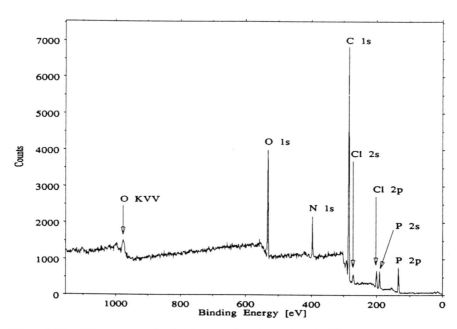

Figure 2.12 XPS spectrum of poly(phenoxyphosphazene) excited by monochromated Al $K\alpha$ (Beamson & Briggs, 1992a).

In the case of a conducting sample in electrical contact with the spectrometer, conservation of energy leads to the equation

$$E_K = h\nu - E_B^F - \phi_{sp} \tag{2.6}$$

where E_K is the measured kinetic energy of the emitted electron, $h\nu$ is the energy of the exciting X-ray photon, E_B^F is the electron binding energy in the conducting solid relative to the Fermi level (see Section 2.1.1.2) and ϕ_{sp} is the work function of the spectrometer ($\sim 5\,\mathrm{eV}$). Since it is the binding energy which is the important physical parameter, the spectrometer is set up to present this automatically as the energy axis, rather than the measured kinetic energy. It is essential that the energy scale is accurately linear and correctly calibrated; this is now a relatively simple matter using clean metal standards and traceable peak energies (Seah and Smith, 1990).

Equation (2.6) refers to the *elastic* photoemission process, i.e. the emitted photoelectrons have suffered no energy loss during transport through the solid. Therefore, for each photon energy in the X-ray source distribution there will be a photoelectron spectrum described by equation (2.6). Thus Fig. 2.11 shows weak 'satellite' peaks; these can be identified from Table 2.1 and eliminated by use of monochromated Al $K\alpha$ radiation. *Inelastic* photoemission is the result of energy loss during transport through the solid. These losses may be discrete, giving rise to weak characteristic features, or due to random scattering. The latter gives rise to the step-like increase in the background on the low kinetic energy (high binding energy) side of each peak in the spectrum. There is also a general contribution to the background from the continuous Bremsstrahlung radiation (for unmonochromated sources) and this is the only contribution at zero binding energy.

The peaks in the spectrum shown in Fig. 2.11 can be grouped into three classes; those due to photoemission from *core levels*, those due to photoemission from *valence levels* and those due to X-ray excited Auger emission (*Auger series*). These will be discussed in much greater detail later, for the moment it is sufficient to note the essential points.

2.2.1.1 Core levels

The peaks in Fig. 2.11 are a direct reflection of the electron shell structure of the silver atom in so far as the Mg $K\alpha$ photons are capable of probing (beyond the 3s level the Ag 2p electrons have a binding energy greater than 1254 eV). The non-s-levels are doublets as a result of spin–orbit coupling. Core-level notation is nl_j where n is the principal quantum number, l is the orbital angular momentum quantum number and j is the total angular momentum quantum number; $j = (l+s)$ where s is the spin angular momentum quantum number ($\pm\frac{1}{2}$). Thus when $l > 0$ there are two possible states (see Table 2.2), the difference in energy of which reflects the 'parallel' or 'anti-parallel' nature of the spin and orbital angular momentum vectors of the remaining unpaired electron following photoemission. Clearly this can be many electron volts. The relative intensities of these doublets are given by the ratio of their respective degeneracies $(2j+1)$, Table 2.2.

With Mg $K\alpha$ and Al $K\alpha$ photons at least one core level is excited for any atom in the periodic table except H (1s electron always in bonding orbitals) and the characteristic binding energy values allow unambiguous elemental identification. Peak overlaps are surprisingly rare. Thus the presence of several elements in the surface layers of the polymer whose XPS spectrum appears in Fig. 2.12 is readily deduced.

The basic parameter which determines the relative intensities of the core level peaks is the atomic photoemision cross-section σ, which depends on the quantum numbers n, l and j and on the proximity of the photon energy to the photoemission threshold. However, for Mg $K\alpha$ and Al $K\alpha$ the *relative* cross-sections for any two levels are very similar. The analyser transmission function and instrumental geometry also affect the relative peak intensities. These are discussed in detail by Seah (1990) and in Section 2.2.3 in relation to quantification.

Figs. 2.11 and 2.12 also clearly show the variation in width of the observed core level peaks. The peak width defined as the fwhm, ΔE, is a convolution of several contributions, so that:

$$\Delta E = (\Delta E_n^2 + \Delta E_p^2 + \Delta E_a^2)^{1/2} \tag{2.7}$$

where ΔE_n is the natural or inherent width of the core level, ΔE_p is the width of the photon source (X-ray line) and ΔE_a is the analyser resolution, all expressed as fwhm. This addition in quadrature assumes that all components have a Gaussian lineshape, which is only an approximation. ΔE_p and ΔE_a were discussed in Sections 2.1.3–2.1.5. ΔE_p is at a minimum when using monochromated Al $K\alpha$ radiation. ΔE_a is constant across the spectrum when the analyser is operated in the CAE mode, but varies when it is operated in CRR mode (constant $\Delta E/E$).

Table 2.2 *Spin–orbit splitting parameters*

Subshell	j values	Area ratio
s	$\frac{1}{2}$	—
p	$\frac{1}{2}, \frac{3}{2}$	1:2
d	$\frac{3}{2}, \frac{5}{2}$	2:3
f	$\frac{5}{2}, \frac{7}{2}$	3:4

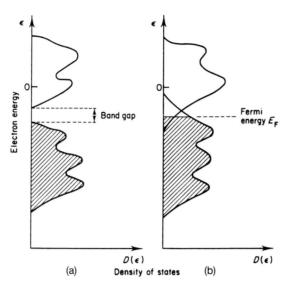

Figure 2.13 Schematic density of states for: (a) an insulator and (b) a metal. The shading indicates the extent to which the energy levels are occupied (after Orchard, 1977).

The natural linewidth is a direct reflection of uncertainty in the lifetime of the core-hole produced by emission of the photoelectron in question. From the Uncertainty Principle, the linewidth is obtained as:

$$\text{linewidth} = \frac{h}{\tau} = \frac{4.1 \times 10^{-15}}{\tau} \text{ (in eV)} \tag{2.8}$$

where Planck's constant (h) is expressed in electron volt seconds and the lifetime (τ) in seconds.

The core-hole lifetime is, in turn, governed by the processes that follow photoemission, in which the excess energy of the ion state decays. The three decay mechanisms are emission of:

(i) an X-ray photon (X-ray fluorescence, XRF), which is rare for ions with excess energy $<3\,\text{keV}$;
(ii) a second electron yielding a double-charged ion via an Auger process; or
(iii) a second electron yielding a doubly-charged ion via a Coster–Kronig process.

These are discussed further in Section 2.2.1.3.

2.2.1.2 Valence levels

Valence levels are those occupied by electrons of low binding energy (say 0–15 eV) which are involved in delocalised or bonding orbitals. Here the spectrum consists of many closely spaced levels giving rise to a *band* structure. This region is usually referred to as the 'valence band'. Two situations need to be distinguished, namely conductors and insulators. Fig. 2.13 illustrates the density of

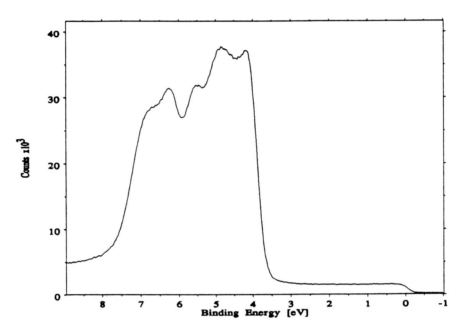

Figure 2.14   Valence band spectrum of clean silver excited by monochromated Al $K\alpha$ (Beamson & Briggs, 1992a).

electron states (per unit energy in unit volume) in these two cases. With insulators the occupied *valence band* is separated from the empty *conduction band*, whilst for metallic conductors these bands overlap and the highest occupied state is termed the Fermi level, E_F. Fig. 2.14 shows the spectrum in this region for silver (labelled Ag 4d in Fig. 2.11) in which the Fermi level cut-off is directly observed. By convention $E_F=0\,eV$ binding energy for energy scale referencing purposes.

In the case of insulating polymers the valence band spectrum is a direct reflection of the occupied density of states but weighted by the photoemission cross-sections of the molecular orbitals involved. The lowest binding energy edge of the spectrum is not, however, E_F. The Fermi level in non-conductors is ill-defined and generally assumed to be located in the middle of the band gap (Fig. 2.13). This poses an energy referencing problem which is discussed fully in Section 3.2.

In general, the valence band is of low intensity because the levels involved, being of low energy, are far removed from the exciting photon energy (and therefore have a low cross-section). Typically, in polymers, the valence band is 20–100 times less intense that the major core lines.

2.2.1.3 Auger series

Auger series are the result of one of the decay mechanisms for the core-hold created during photoemission, shown in Fig. 2.15. For nearly all of the elements associated with polymers this mechanism dominates over X-ray fluorescence.

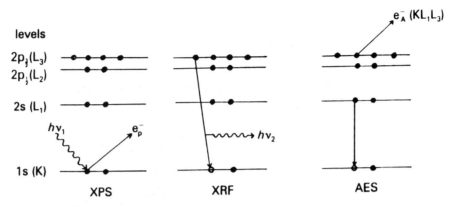

Figure 2.15   Photoemission and subsequent relaxation processes (XRF and Auger electron emission) in XPS.

The Auger process involves three steps: first, the creation of a core hole (in AES this is achieved by irradiation with 5–10 keV electrons); second, the filling of this hole by an electron dropping from a less tightly bound level; third, the emission of an (Auger) electron which takes up the remaining excess energy as kinetic energy. If the levels involved are, respectively, of energy E_1, E_2 and E_3 then the kinetic energy of the Auger electron is given, to a first approximation, by

$$E_K^A = E_1 - E_2 - E_3 \tag{2.9}$$

and since all three levels are characteristic for the atom involved, E_K^A is likewise characteristic and element specific. Because some of the levels involved can be quite close in energy, there are several competing processes which gives rise to a series of peaks (hence, Auger series) in a particular region of the 'photoelectron spectrum'.

The nomenclature for Auger peaks uses 'physics' notation, historically associated with X-ray spectroscopy, rather than the 'chemistry' notation used for core levels and in spectroscopy generally. Thus quantum numbers $n=1,2,3,4\ldots$ are designated $K,L,M,N\ldots$ respectively, whilst l_j states are given suffixes: e.g. (K=1s, L_1=2s, L_2=$2p_{1/2}$, L_3=$2p_{3/2}$, M_2=$3p_{1/2}$, M_3=$3p_{3/2}$, M_4=$3d_{3/2}$, M_5=$3d_{5/2}$ etc). The Auger peak is designated by the three levels involved, e.g. KL_1L_3. Often one or two of the levels involved will actually be in the valance band and this is sometimes referred to as V. So for carbon, for instance, where the L levels are involved in bonding, the alternative term would be KVV (Fig. 2.12). The Coster–Kronig process is similar to the Auger process except that the final doubly-ionised state includes one hole in a shell with the same principal quantum number as that of the original core-hole, e.g. $L_1L_2M_4$. These transitions give rise to very low energy emitted electrons which are not normally detected. Also, they are very fast for 2s and 3s core-holes which leads to broad photoelectron peaks (e.g. Ag 3s in Fig. 2.11).

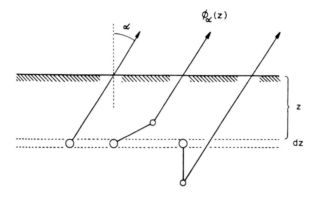

Figure 2.16 Schematic showing (from left to right) representative trajectories with no elastic scattering, a single small angle, elastic scattering, and a single, large scale elastic scattering that contribute to the depth distribution function $\phi_\alpha(z)$ for a particular emission angle α and a particular depth z (Jablonski & Powell, 1993).

2.2.2   The sampling depth and related parameters

As noted above, the surface sensitivity of XPS is not related to the penetration depth of the exciting X-rays, which is many microns. Rather, it is due to the low probability that photoelectrons generated just below the surface will leave the solid with their original energy and contribute to the peaks which are described, in terms of this energy, by equation (2.6). The depth of origin of these electrons is therefore of fundamental importance. Unfortunately, our understanding of this subject is still developing and the situation is further complicated by the existence of a number of terms in the literature which, although defined differently, have been (and still are) used interchangeably. The following attempt at clarification must necessarily be rather brief and the review by Jablonski & Powell (1993) is recommended for greater detail.

The possibility that a photoelectron generated at some point below the surface will leave the solid with its original energy is clearly determined, at least partly, by processes which lead to energy loss during transport through the solid (inelastic scattering processes). A number of terms for describing inelastic scattering effects have been used in the literature: the inelastic mean free path, the attenutation length and the escape depth. These terms are defined and it is important to note that they were introduced whilst it was generally believed that only inelastic scattering was important. During the last 5 years it has been realised that elastic scattering is also important and this has greatly complicated the situation.

Consider Fig. 2.16. Photoelectrons are created at a depth z below the surface with a certan angular distribution depending on the orbital involved (see Section 2.2.3.2). For the case of the straight-line trajectory only inelastic scattering leads to electron attenuation, which is assumed to obey the same experimental formula as does light attenuation in an absorbing medium (the Beer–Lambert law), thus:

$$I_z = I_0 \exp\left(-z/\lambda \sin\theta\right) \tag{2.10}$$

where I_z is the intensity emanating from the atoms at depth z, I_0 is the intensity from the surface atoms and θ is the electron 'take' off angle to the surface ($\theta = 90 - \alpha$). In this case λ is the inelastic mean free path of the measured electrons, defined as the average distance that an electron with a given energy travels between successive inelastic collisions (ASTM E42 1991).

For a sample consisting of an overlayer of material A, thickness d, and a bulk material B (where A and B are both of uniform composition) equation (2.10) leads to be expressions.

$$I_A = I_A^\infty[1 - \exp(-d/\lambda_A \sin \theta)] \tag{2.11}$$

$$I_B = I_B^\infty[\exp(-d/\lambda_B \sin \theta)] \tag{2.12}$$

where I_A and I_B are the measured intensities, I_A^∞ and I_B^∞ are the intensities from bulk A and B measured under identical conditions, λ_A and λ_B are the inelastic mean free paths for electrons of the measured core-electrons of A and B, respectively, travelling through material A (in general these will differ because λ is a function of kinetic energy as discussed later).

These equations have formed the basis for experimental determinations of λ, in which overlayers have been deposited onto substrates with some independent measurement of film thickness. However, in this case λ is not actually the inelastic mean free path. Elastic scattering events, as depicted in Fig. 2.16, lead to longer trajectories compared to the straight-line trajectory (for inelastic scattering only). According to theoretical calculations this will lead to λ values less than the inelastic mean free path by 30% or more. Knowledge of λ is important for any quantitative XPS measurement: overlayer thickness, average depth of analysis and (perhaps most important) surface composition (see Section 2.2.3). Because the λ values used to date have been predominantly those derived from the results of the 'attenuation' experiments they are now frequently referred to as attenuation lengths, although there is no currently accepted definition of attenuation length.

Use of equations (2.11) and (2.12) implicitly assumes exponential intensity dependences but calculations have shown that, in general, electron attenuation in solids is not exponential. Indeed, many experimental results did not give exponential dependences, but these were discarded on the assumption that the unexpected behaviour was due to film non-uniformities (e.g. island growth of deposited layers). Furthermore, attenuation length values are dependent on the experimental geometry.

These problems have led to the introduction of the depth distribution function, one definition of which is: for electrons leaving a solid at a certain angle α to the surface normal, the depth distribution function describes the contributions ΔI corresponding to electrons emitted within a layer of thickness Δz at a depth z and reaching the surface without energy loss.

It is most likely that in future all the terms discussed in this section will be consistently defined in terms of the depth distribution function. However, returning to the current definitions which assume an exponential decay of contributions to the signal with increasing depth, the *escape depth* is defined as the distance, normal to the surface, at which the possibility of an electron escaping without significant energy loss due to inelastic scattering process dropped to e^{-1} (36.8%) of its original value. 95% of the signal detected comes from a depth of three times the escape depth.

Again, for the case of exponential decay, if the attenuation length is λ_{AL} then for a take-off angle, θ, to the surface, the escape depth, d is $\lambda_{AL} \sin \theta$. Commonly $3d$ (95% detection) is taken to be the 'sampling depth' ($=3\lambda_{AL} \sin \theta$). The variation of this parameter with take-off angle is the basis of 'angular resolved' XPS for non-destructive depth profiling. This is discussed in detail in Section 3.8.

2.2.2.1 Sources of attenuation length and inelastic mean free path values

Experimental values for λ_{AL} up to 1978 were compiled (Seah and Dench, 1979) and fitted by universal curves so that unknown λ_{AL} could be estimated. (In the original publication these data were referred to as inelastic mean free paths). The values were plotted separately for elements, inorganic compounds and organic compounds. The last of these is shown in Fig. 2.17 and the fitted equation is

$$\lambda_{AL} = (10^3/\rho)\,(49E^{-2} + 0.11\,E^{0.5}) \tag{2.13}$$

where λ_{AL} is in nanometres, E is the kinetic energy in electron volts and ρ is the density in kilograms per cubic metre. Clearly the data are limited and also show a great deal of scatter. Subsequent data have not lessened the uncertainty, largely due to the variety of materials used – ranging from plasma-deposited cross-linked polymer films to Langmuir–Blodgett films of fatty acid salts – and the different experimental techniques employed. The issues involved have been discussed in detail (Roberts *et al.*, 1980). These authors studied poly(methyl methacrylate) (PMMA) films on silicon, deposited by spin casting, with thickness variation determined by ellipsometry. These results are probably of greater relevance for studies of conventional polymers, giving λ_{AL} (PMMA) of 2.9 ± 0.4 nm ($E=1196$ eV) and 3.3 ± 0.5 nm ($E=1328$ eV). These values refer to C 1s and Si 2p electrons, respectively, excited by Al $K\alpha$ radiation. Using $\rho = 1.188 \times 10^3$ kg m^{-3} in equation (2.13) gives $\lambda_{AL} = 3.2$ nm ($E=1196$ eV), thus in this case, at least, the agreement is fair.

To be most useful these data should provide the general relationship between λ_{AL} and E for the material in question. This is most frequently expressed as a simple power relationship

$$\lambda_{AL} = KE^x \tag{2.14}$$

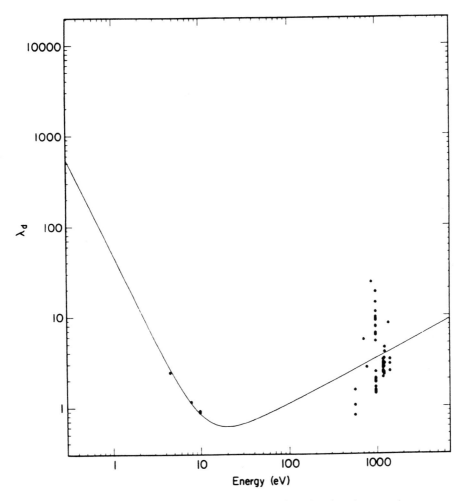

Figure 2.17 Compilation of experimental attenuation length values for organic compounds and fitted 'universal curve' (Seah and Dench, 1979).

For the energy range of most interest (500–1500 eV) the first term in equation (2.13) is negligible compared to the second and $x=0.5$. For a wide range of materials values of x between 0.5 and 0.7 emerge. However, for individual experiments on organic materials values of $x>1$ are common (e.g. $x=1.2$ for the Roberts *et al.* data quoted above) and the scatter is very large.

Inelastic mean free paths can be calculated from experimental optical data. The most complete source of data arises from work by Tanuma and coworkers (for calculations for 14 organic compounds, including several polymers, see Tanuma, Powell & Penn (1994)). The calculations generate inelastic mean free path (λ) values over the energy range 50–2000 eV and these can be fitted by a modified form of the Bethe equation for inelastic scattering:

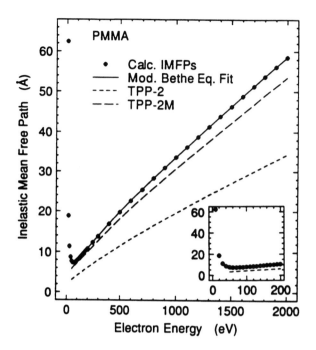

Figure 2.18 Calculated inelastic mean free path values for PMMA as a function of kinetic energy and fitted curves (see text) (Tanuma *et al.*, 1994).

$$\lambda = \frac{E}{E_p^2[\beta\ln(\gamma E) - (C/E) + (D/E^2)]} \tag{2.15}$$

where λ is in angstroms, E is the kinetic energy (in electron volts), $E_p = 28.8(N_v\rho/M)^{0.5}$ is the free-electron plasmon energy (in electron volts), ρ is the density (in grams per cubic centimetre), N_v is the number of valence electrons per molecule and M is the molecular weight. The fitting parameters β, γ, C and D can be empirically related to other material properties as follows:

$$\beta = -0.10 + 0.944/(E_p^2 + E_g^2)^{0.5} + 0.069\rho^{0.1} \tag{2.16a}$$

$$\gamma = 0.191\rho^{-0.5} \tag{2.16b}$$

$$C = 1.97 - 0.91U \tag{2.16c}$$

$$D = 53.4 - 20.8U \tag{2.16d}$$

where $U = N_v\rho/M$ and E_g is the band-gap energy (in electron volts). Equations (2.15) and (2.16) are referred to as TPP-2M (Tanuma, Powell and Penn, second modification). The average rms deviation of the calculated λ values from those obtained from the predictive formula TPP-2M for all 14 materials was 8.5% and the specific results for PMMA are shown in Fig. 2.18. For comparison with the experimental results discussed previously, λ (1200 eV) = 3.88 nm. This is greater than λ_{AL} for the same energy, as expected. Elastic scattering becomes more important as atomic number increases and Seah (1990) has suggested the following relationship, based on other literature data:

$$\lambda_{AL}/\lambda = (1-0.028\,Z^{0.5})(0.501+0.068\,1nE) \tag{2.17}$$

For PMMA (repeat unit $C_5H_8O_2$) the average value of Z is 3.6. At 1200 eV equation (2.17) gives $\lambda_{AL}/\lambda = 0.74$, exactly the figure obtained from the ratio of the experimental ($\lambda_{AL} = 2.9$ nm) and calculated ($\lambda = 3.9$ nm) values discussed above. Despite this result, it is fair to conclude that much work still needs to be done to establish reliable $\lambda_{AL}(E)$ data for polymeric materials.

Tanuma et al. (1994) found that above ~500 eV their calculated inelastic mean free paths for polymers could be well fitted by $\lambda \propto E^{0.79}$. In order to provide a guide to the variation in sampling depth for polymeric materials the data in Table 2.3 have been calculated on the basis of this energy dependence of λ_{AL} and the value of λ_{AL} for C1s in PMMA (excited by Al $K\alpha$) of 2.9 nm obtained by Roberts et al. (1980). The chosen core levels span most of the range of interest in polymer studies.

2.2.3   Quantification of surface atomic composition

Quantification of peak intensity data is performed using peak areas. Consequently the form of the background which needs to be subtracted is of major importance, so this is considered first. Next, the terms which contribute to the measured intensity are discussed so that the simplifications involved in the different approaches to quantification can be appreciated.

Table 2.3   *XPS sampling depths as a function of core level binding energy and take-off angle (measured from the surface plane)*

Core level	Binding energy (eV)	Sampling depth (95% signal) (nm)					
		Mg $K\alpha$			Al $K\alpha$		
		10°	45°	90°	10°	45°	90°
F 1s	686	0.8	3.4	4.8	1.1	4.5	6.3
O 1s	531	1	4.1	5.8	1.3	5.2	7.3
N 1s	402	1.1	4.7	6.6	1.4	5.7	8
C 1s	287	1.3	5.2	7.3	1.5	6.2	8.7
Si 2p	102	1.5	5.9	8.4	1.7	6.9	9.7

2.2.3.1 Background subtraction

Three approaches to defining the spectral background are discussed in the literature: the straight line, the Shirley method and the Tougaard method. The first two are commonly available on most (even old!) datasystems whilst the latter is much more recent. A detailed description of the relative merits of the three approaches has been given (for metallic systems only) by Tougaard & Jansson (1993). In the following analytical expressions J is the measured spectrum (strictly following correction for the analyser transmission function, see Section 2.2.3.2), F is the background-corrected spectrum over the chosen kinetic energy region $E_{min} < E < E_{max}$ with corresponding channels (width ΔE) $i_{min} < j < i_{max}$. For the straight line

$$F(i) = J(i) - k \sum_{j=i+1}^{i_{max}} \Delta E \qquad (2.18)$$

where k is found by the requirement that $F(i_{min}) = 0$. For the Shirley method

$$F_n(i) = J(i) - k_n \sum_{j=i+1}^{i_{max}} F_{n-1}(j) \, \Delta E \qquad (2.19)$$

where k_n is found by the requirement that $F_n(i_{min}) = 0$. This requires iteration but the series typically converges to $F_n \sim F_{n-1}$ within five iterations. For the Tougaard method

$$F(i) = J(i) - B_1 \sum_{j=i+1}^{i_{max}} \frac{(j-i) \, \Delta E}{[C + (j-i)^2] \, \Delta E^2} J(j) \, \Delta E \qquad (2.20)$$

where $C = 1643 \, eV$. B_1 is a parameter adjusted to give $F \sim 0$ in a wide energy range below that of the real structure (typically 50 eV). The choice of E_{max} is usually straightforward, i.e. within a few eV of the onset of the peak (on the high kinetic energy side). The choice of E_{min} is often more difficult, especially when shake-up structure is involved (see section 3.5), or when a range of chemically shifted components 'fills in' the gap between the core line and the onset of the steeply sloping loss structure. Since the Shirley method assumes that each unscattered electron is associated with a flat background of losses, the background intensity at a point is proportional to the intensity of the total peak area (above background) at any point. However, the discrete gap between peak and losses seen for polymers reflects the band gap associated with insulators generally; therefore the validity of applying the Shirley method to polymer spectra is questionable although its use in the literature is widespread (see Fig. 2.19).

The derivation of the Tougaard algorithm is beyond the scope of this book. It appears to be very successful for treating spectra from metallic systems, revealing the large fraction of the 'real' peak area usually ignored by the other methods necessarily applied over a much smaller energy range. However, its applicability to polymer spectra is as yet unknown and its availability on datasystems very restricted.

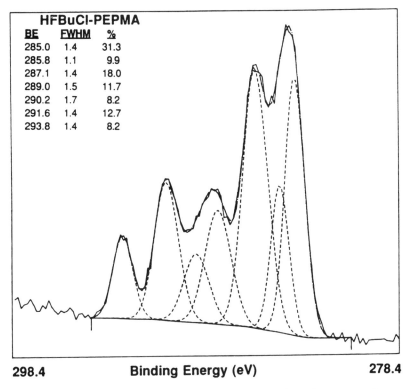

Figure 2.19 Example of Shirley background used in the C1s region of a spectrum from a complex polymer (Chilkoti & Ratner, 1991).

For quantitative analysis of polymer spectra, therefore, the use of the straight line background is recommended provided some care is taken to ensure consistent application (i.e. choice of end-points).

2.2.3.2 Factors involved in XPS intensities

If an incident X-ray photon of energy $h\nu$ ionises core level X of element A in a solid, the photoelectron current detected by the spectrometer is

$$I_A(X) = I_0 p \sigma (h\nu, E_X) L(h\nu, X) G(E_X) D(E_X) \int_0^\infty N_a(z) \exp\{-z/\lambda_m(E_X)\cos\alpha\} dz \qquad (2.21)$$

where I_0 is the X-ray flux, illuminating the sample, p is a surface roughness factor (which affects X-ray illumination and photoelectron ejection through shadowing effects), $\sigma(h\nu, E_X)$ is the photoionisation cross-section for ionisation of X by photon $h\nu$, $L(h\nu, X)$ is the angular asymmetry factor for emission from X by photon $h\nu$, $G(E_X)$ is the spectrometer étendue (the product of the transmission efficiency and the area from which the electrons are emitted) at kinetic energy E_X, $D(E_X)$ is the detector efficiency at energy E_X, $N_a(z)$ is the distribution of atoms A with depth z, $\lambda_X(E_X)$ is the inelastic mean free path of electrons

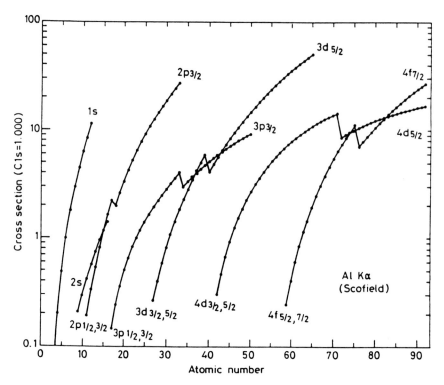

Figure 2.20   Calculated values of the cross-section (σ) for Al $K\alpha$ radiation normalised to σ (C 1s)=1 (Seah (1990) using original data from Schofield (1976)).

E_X in matrix M (containing atoms A) and α is the angle of emission (to the surface normal).

If it can be assumed that the X-ray flux is constant and that the analysed volume is of uniform composition, then equation (2.21) can be simplified to

$$I_A(X)=K\sigma LN_A\lambda_m\cos(\alpha)GD \tag{2.22}$$

which describes the factors affecting the measured peak intensity. Terms relating to the excitation process are σ and L. The photoionisation cross-section, σ, is the transition probability per unit time for excitation of a single photoelectron (A_x) under an incident photon flux of L $cm^{-2}s^{-1}$. σ depends on $h\nu$, atomic number (Z) and core level (n, l). Values calculated by Schofield (1976) are shown in Fig. 2.20. Whilst these are for Al $K\alpha$ excitation, the *relative* values for Mg $K\alpha$ excitation are very similar. σ also depends on the angle between the incident photon direction and the direction of photoelectron detection (γ) and this is embodied in the angular asymmetry factor, L_A

$$L_A=1+\tfrac{1}{2}\beta_A(\tfrac{3}{2}\sin^2\gamma-1) \tag{2.23}$$

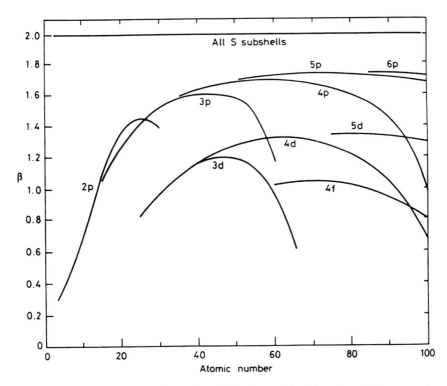

Figure 2.21 Calculated values of β (Seah (1990) using original data from Reilman *et al.* (1976)).

where β_A, the asymmetry parameter, is a constant for any given atom/sub-shell/photon combination. Values of β have been calculated for the common X-ray energies, as illustrated in Fig. 2.21 (Reilman, Msezane & Manson, 1976). Whilst these values agree well with data for gases, elastic scattering effects in solids (discussed above) reduces the effect of β.

Thus for normal emission, Seah (1990) suggests that the calculated values be reduced according to the following dependence on the average atomic number (Z):

$$\beta^* = \beta(0.781 - 0.00514\ Z + 0.000021\ Z^2)$$

Therefore it is more appropriate to replace L_A and β in equation (2.23) by L_A^* and β^*. Note also that for $\gamma = 54.74°$, L_A does not depend on γ and for many of the more recent instrument designs γ is within 10° of this 'magic angle'.

The spectrometer-dependent terms in equation (2.22) are G and D. The spectrometer étendue, G, depends on kinetic energy E_K, often in a complex way, and is a function of input lens set-up (magnification, aperture/slit dimensions), analyser type/operation and the X-ray illumination (large area flood, or monochromated focused spot). The expected relations for $G \propto E_k^p$ are derived by Seah (1989, 1990) from electron-optical first principles. For operation in the constant $\Delta E/E$ mode, the analysed area is constant and $G \propto E_K$. For operation in the constant ΔE mode the transmission curve typically moves from a $G \propto E_K^{-0.5}$ to $G \propto E_K^{-1}$ dependence as E_K increases, the transition point for any set of otherwise constant conditions depending on ΔE (see also Section 2.1.5).

The effect of detector efficiency, which can be markedly dependent on E_K, can be ignored when operating at constant ΔE since all electrons are detected at this energy. In the constant $\Delta E/E$ mode the detection energy will vary with E_K/R ($R=E/\Delta E$) and this could mean a variation in D of a factor of 2 or more over the range $100 < E_K < 1500\,\mathrm{eV}$ for a typical channeltron (Seah, 1990). The change in detector performance with time and the variation in individual detectors are additional reasons why constant ΔE operation is preferred for quantitative work.

2.2.3.3 Relative sensitivity factors

From equation (2.22) it is clear that $I_A(X) \propto N_A$ under fixed experimental conditions, i.e. the detected photoelectron current is linearly related to the atom density which we want to determine. However, since the proportionality constant contains a roughness term it would be very difficult, even if possible in principle, to derive N_A knowing $I_A^\infty(X)$ – the intensity from a sample of pure A. Therefore all XPS data are quoted in terms of atom fractions (the percentage of all atoms detected – note that hydrogen is excluded). This is achieved through the use of relative sensitivity factors, with one elemental peak taken as the standard to which other peaks are referred (usually F 1s). Thus for a stoichiometric compound containing fluorine, from equation (2.22):

$$\frac{I_A}{I_F} = \frac{K\sigma_A^* \, N_A \lambda_A \cos(\alpha) \, G_A \, D_A}{K\sigma_F^* \, N_F \, \lambda_F \cos(\alpha) \, G_F \, D_F} \tag{2.24}$$

where $\sigma^* = \sigma L^*$ (the differential photoemission cross-section) and the subscripts A and F refer to level X of atom A and F 1s (specifically their particular energies), respectively. Defining $I_F(1s)=1.00$ and cancelling identical terms gives:

$$I_A = \frac{\sigma_A^* \, \lambda_A \, G_A \, D_A}{\sigma_F^* \, N_F \, \lambda_F \, G_F \, D_F} \, N_A = S_A N_A \tag{2.25}$$

where S_A is the relative sensitivity factor of A (measured using level X).

Relative sensitivity factors can be calculated. The factors required to compute σ^* values have been described, as has the energy dependence of λ. Data for the energy dependence of G (often simply referred to as the instrument transmission function) is now available for many instruments (Seah, 1993) and D is usually a constant (certainly, if operating in CAE mode). In much early work using CAE (constant ΔE) analyser operation it was assumed that since $\lambda_A \sim E^{0.5}$ and $G \sim E^{-0.5}$, then λG could be treated as a constant and $S_A = \sigma_A^*$.

Alternatively, relative sensitivity factors can be determined empirically. Several tabulations exist of which the most generally useful is that due to Wagner *et al.* (1991). These data were obtained from freshly powdered stoichiometric compounds mainly containing fluorine, and therefore allowing direct reference to I_F, on instruments claimed to have $G \propto E^{-1}$ when operated in the constant ΔE mode. An inherent problem is the presence of absorbed 'hydrocarbon contamination' so that all intensities (I_A) are reduced by the factor exp $\{-t/\lambda_c(E_A)\sin \theta\}$, where t is the contamination thickness – typically 0.75 nm – and θ is the take-off angle with respect to the surface. For use with quantification of samples not so contaminated (polymers are likely to be in this category) a correction needs to be applied. The low kinetic energy of F1s exacerbates this particular problem. By comparing ratios of (corrected) experimental and calculated relative sensitivity factors Seah (1990) has concluded that for peak intensities measured using the restricted range background subtraction methods (linear, Shirley), quantification is more accurate when empirial relative sensitivity factors are employed as compared with the calculated values. Probably the main reason for this is the fact that the measured intensity omits shake-up and shake-off events (even allowing for the inclusion of obvious shake-up structure, see Section 3.5). Use of the Tougaard background subtraction algorithm readily allows the magnitude of this omission – which may be as high as 30% – to be assessed. The calculated relative sensitivity factors, being based on calculated photoionisation cross-sections, implicitly include all these events.

Use of Wagner relative sensitivity values with any accuracy is only possible if either the instrument used has the same transmission function (E^{-1}) as that of the instruments used for their determination or the difference in transmission functions is allowed for. For this reason it may be most expedient to determine relative sensitivity factors under the particular instrument conditions routinely used. The polymers should be chosen carefully to minimise the complications due to shake-up noted above (i.e. fully saturated aliphatic polymers are preferable) and to maximise the chances of providing fresh, clean samples by spin casting onto a low Z substrate such as silicon, aluminium or glass (see Section 3.3). Table 2.4 gives some suggestions for suitable standards. It is most convenient to use C1s=1 (rather than F1s=1) in a set of internally generated relative sensitivity factors for use in polymer surface analysis.

With these relative sensitivity factors (S_n) the relative atomic concentration of any chosen element, A, is then simply obtained from

$$C_A = \frac{I_A S_A}{\Sigma_n (I_n / S_n)} \tag{2.26}$$

where C_A is usually expressed as atomic % (of all elements determined, hydrogen excluded).

Table 2.4 *Recommended polymers for the measurement of atomic relative sensitivity factors*

Element	Polymer	Repeat unit	Casting solvent
O	poly(propylene glycol)	$-CH-CH_2-O-$ with CH_3 on CH	chloroform
O	poly(vinyl methylether)	$-CH_2-CH-$ with OCH_3	toluene
N	poly(methacrylonitrile)	$-CH_2-C-$ with CH_3 and $C{\equiv}N$	acetone
N	poly(allylamine)	$-CH_2-CH-$ with CH_2NH_2	water
O/N	poly(2-ethyl-2-oxazoline)	ring: $O-CH_2$, C, CH_2, N, CH_3CH_2	chloroform
N/Cl	poly(allylamine hydrochloride)	$-CH_2-CH-$ with $CH_2\overset{+}{N}H_3\ Cl^-$	water
Cl	poly(vinyl chloride)	$-CH_2-CH-$ with Cl	THF
F	poly(tetrafluoroethylene)	$-CF_2-CF_2-$	–
F/Cl	poly(ethylene-co-chloro trifloroethylene)	$-CH_2-CH_2-CF-CF_2-$ with Cl	–
S	poly(ethylene sulphide)	$-CH_2-CH_2-S-$	–
Si/O	poly(dimethylsiloxane)	$-Si-O-$ with CH_3 and CH_3	hexane
Br	poly(4-bromostyrene)	$-CH_2-CH-$ with phenyl–Br	toluene

Chapter 3

Information from polymer XPS

Before discussing the interpretation of polymer spectra it is important to consider two factors which may cause problems. These are sample charging, with the associated problem of binding energy scale referencing, and radiation damage.

3.1 Sample charging

Emission of electrons from the surface of an insulator during X-irradiation leads to the creation of a positive surface potential since electrons from the bulk of the sample or from the sample mount cannot compensates for this loss. The effect of this positive charging is a decrease in the kinetic energy of the photoelectrons (i.e. an apparent increase in binding energy). In the case of non-monochromatic X-ray sources a compensating flux of electrons is fortuitously provided by the window of the X-ray source (generally a thin aluminium or beryllium foil) and possibly by impingement of the broad X-ray beam onto other metallic components in the sample vicinity. The overall result is a small equilibrium potential, which is effectively established instantaneously, of typically <10 eV (dependent on instrument configuration, take-off angle and X-ray source operating conditions for a given polymer).

In the case of monochromated X-ray sources the situation is much more serious. Firstly, the X-rays are generated remote from the sample. This means that a source of compensating electrons has to be introduced deliberately, generally from a high current, low energy electron source. This is usually referred to as an electron 'flood' gun because the beam is designed to cover a large area.

Secondly, the X-rays are focused onto a small area and the lack of uniform irradiation can lead to *differential charging* of the sample. Furthermore, unless the X-ray and electron beams are close to coaxial there will be the additional problem that effective neutralisation may only be possible over a restricted range of take-off angle. As will be seen below, however, the benefits of using mono-chromated radiation are sufficiently great that there has been a major drive to overcome these problems (and to understand charging in more detail). It is par-ticularly advantageous to be able to 'tune' the surface potential by varying para-meters and visualising peak shape/intensity in real time. With the advent of very high sensitivity instruments employing efficient monochromators, this is now possible (Beamson & Briggs, 1992a). The best strategy seems to be to drive the surface potential slightly negative (by ~2–3 eV).

Since photoelectron emission is a key component of charging it would be anticipated that the extent of charging (i.e. the equilibrium potential established under certain experimental conditions) is dependent on polymer surface composition. Thus materials with low photoemission cross-sections (e.g. poly(ethylene), which contains only carbon and hydrogen) charge to a much lesser extent than high cross-section materials (e.g. PTFE, which contains fluo-rine with a photoemission cross-section four times that of carbon). A useful account of some of the systematic effects involved in charging of polymers under XPS conditions can be found in Dilks (1981).

Polymer charging can be avoided altogether by studying very thin films spun-cast from dilute solution onto a conducting substrate. Provided that the film thickness is comparable to the electron escape depth (i.e. ~2.5 nm), secondary electrons generated in the substrate can pass through the film and compensate incipient charging. In order to avoid radiation damage problems, however, the substrate needs to be chosen with care (see Section 3.3). Silicon wafer, ion etched just prior to spin-casting (to remove organic contamination) is suitable. This is an ideal way to establish the minimum line-widths for standard samples prior to optimising charge neutralisation strategies for analysing the surfaces of nor-mally insulating samples.

3.2   Binding energy scale referencing

It was noted in Section 2.2.1.2 that for conducting samples, which can be mounted so as to be in electrical contact with the spectrometer, it is easy to estab-lish the position of zero binding energy as the Fermi level, E_F. It was also noted that for insulators, in general, there is no such measurable spectral feature since E_F is ill defined and above the measured valence band edge. Therefore binding energy referencing for polymers is a requirement, but variable charging means that this process needs to be carried out on an individual basis for each spectrum.

In principle the most satisfactory procedure is that of internal referencing, by which a particular peak in the spectrum can be unambiguously assigned to an accurate binding energy (E_B). If in the measured spectrum, this peak appears at $E_B \pm \delta$ eV, then all the other peak energies are corrected appropriately for the charging shift ($\pm \delta$). Alternatively, the datasystem software shifts the whole binding energy scale by this amount. A common correction is to put saturated hydrocarbon C1s at 285.00 eV. This is convenient since many polymers contain such units either in the polymer backbone or in a side chain. It was believed that all carbon atoms bound only to themselves and/or hydrogen atoms, irrespective of hybridisation, had this same C1s binding energy. As will be shown below this is actually not the case and a secondary standard of C1s for unfunctionalised aromatic carbons of 284.70 eV has been suggested (Beamson & Briggs, 1992a).

There will be instances when either these common reference points are not available or there is insufficient evidence to assign a peak to one of them. In such instances an external reference must be used. In the pioneering early work of Clark *et al.* (see Dilks, 1981) the build up of 'hydrocarbon' on a polymer surface with time provided such a reference (through a growing C1s 285.00 eV peak), although it is not clear that the adsorbing material was characterised. The instrument used for this work was non-UHV and such working conditions are unlikely to be relevant to modern UHV systems. The equivalent effect can be achieved by deliberate external contamination with hydrocarbon. The polymer surface is contaminated by gently rubbing a few crystals of the saturated linear hydrocarbon hexatriacontane (n-$C_{36}H_{74}$) over it, using a clean spatula, and blowing off the excess. As an example, Fig. 3.1 shows the C1s spectra of poly(vinyl chloride), which has the repeat unit $-CH_2-CHCl-$, before and after this treatment. The degree of contamination indicated is sufficient to established an accurate position for C1s=285.00 eV without causing any change to the magnitude of the original charging effect. Quite small binding energy shifts can be measured in this way, e.g. C1s 285.52 eV for the single peak of poly(ethylene sulphide), $(CH_2CH_2S)_n$ (Beamson & Briggs, 1992a).

3.3 Radiation damage

Although XPS is a relatively benign technique in terms of sample degradation induced by the exciting beam, polymers are not immune to radiation damage during spectra acquisition. It is also commonly believed that any problems are reduced when using a monochromated source of X-rays, because of the absence of Bremsstrahlung radiation. However, this is probably an oversimplification. Most non-monochromated sources are placed very close to the sample and thermal effects are also likely, since temperature rises of several tens of degrees

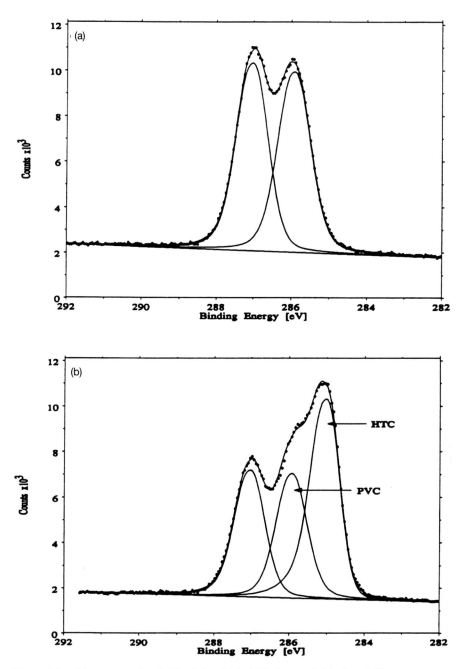

Figure 3.1   C1s spectra of poly(vinyl chloride): (a) before and (b) after deliberate contamination with hexatriacontane (Beamson & Briggs, 1992a).

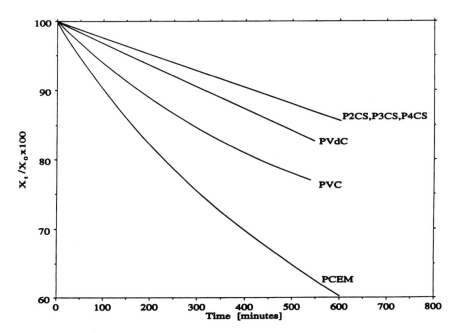

Figure 3.2 Plot of $(X_t/X_0) \times 100$ vs t for chlorine-containing polymers, where X=Cl/C atom ratio: poly (2-,3- and 4-chlorostyrene), poly(vinylidene chloride), poly(vinyl chloride) and poly(chloroethyl methacrylate) (Beamson & Briggs, 1992a).

Celsius can result from radiative heating by the hot X-ray gun casing. Sample cooling can reduce these effects (and the loss of volatile additives which may be present – a particular problem in studies of formulated rubber systems).

Detailed *in situ* degradation studies of a number of polymers have been reported (see Briggs (1990b)) but these typically involve exposure times much longer than normally required for analysis. Of more practical interest is the relative stability of a particular polymer, or class of polymer. This allows some judgement to be made about acquisition times particularly when these need to be relatively long (e.g. in image acquisition or high resolution studies of weak spectral features). Such a database for over 100 homopolymers has been compiled in which a simple degradation index, measured under constant operating conditions, is quoted. This index was estimated from a graph of X_t/X_0 vs t, where X is a parameter derived from the XPS spectrum which is characteristic of the undamaged polymer (e.g. the Cl/C atomic ratio for a chlorine-containing polymer); X_t is its value at time t of exposure to the X-rays (in this case monochromated Al$K\alpha$) and X_0 its value at t=0. In most cases X is related to atomic composition, as illustrated in Fig. 3.2 (Beamson & Briggs, 1992a). The X-ray degradation index quoted is the percentage fall in X_t/X_0 after 500 min exposure at a source power of 1.4 kW, rounded to the nearest 5%. A set of representative values for comparison are given in Table 3.1.

It is worth noting that this index will not indicate a change in structure not accompanied by heteroatom loss (if an atom ratio is used to describe X). Also it is known that rapid depolymerisation (which does not change the surface stoichiometry) can occur, giving rise to volatile species detectable by a mass spectrometer attached to the XPS chamber.

Electrons generated in the sample are very important in the initiation of degradation mechanisms. This can be most noticeable when studying very thin films on conducting substrates (as mentioned above), a common technique for investigating polymer–metal interactions. Degradation rates can be markedly enhanced when the metal has a high photoemission cross-section (e.g. gold) leading to a high flux of photoelectrons passing through the polymer film.

3.4   Core levels

Core level spectra provide most of the information derived from polymer XPS. As discussed in Chapter 2, core level binding energies and relative intensities provide atomic identification and relative concentrations, respectively. For any given atom, however, the exact core level binding energy varies slightly

Table 3.1   *Degradation indices for a range of polymers (see text for the definition of the parameters)*

Polymer	X	Degradation index
poly(propylene)	C1s FWHM	5
poly(cis-butadiene)	shake-up/C1s	60
poly(styrene)	shake-up/C1s	0
poly(ethylene glycol)	O/C	5
poly(ethyl acrylate)	O/C	10
poly(methyl methacrylate)	O/C	10
cellulose	O/C	10
poly(ether ether ketone)	O/C	0
poly(acrylamide)	N/C	10
cellulose trinitrate	N/C	65
poly(tetrafluoroethylene)	F/C	10
poly(vinyl chloride)	Cl/C	25
poly(2-chloroethyl methacrylate)	Cl/C	35
poly(dimethyl siloxane)	C/Si	15

depending on its local bonding situation – producing so-called binding energy or 'chemical' shifts. Unfortunately the range of these shifts is small (~10 eV maximum in the case of C 1s and N 1s but generally even less) and therefore the dynamic range, i.e. the binding energy range relative to the peak width (measured as the fwhm), is very low and critically dependent on the energy resolution. Consequently, when the polymer structure involves the same atom in more than one chemical environment, then the resulting chemically shifted peaks will probably overlap to produce a complex envelope. This problem is greatest, of course, in the C 1s case. Extraction of structural information from such spectra requires the component peaks to be resolved and this, in turn, requires knowledge of individual component lineshapes.

3.4.1 Lineshapes

As discussed in Section 2.2.1.1 there are several contributions to the photoelectron line-width, but the associated lineshapes are not the same. The $K\alpha_{1,2}$ photon source lineshape is essentially Lorentzian, whilst the overall contribution from the spectrometer is essentially Gaussian. The natural core level lineshape is described by the Doniach–Sunjic theory and is essentially Lorentzian with a small degree of asymmetry ('tailing') to high binding energy (Riviere, 1990b). In addition there are significant solid-state broadening effects due to variation in local structure experienced by any one 'type' of atom (e.g. polymer chains in the surface layer relative to those below, crystalline relative to amorphous chain sections). This is due to variation of intermolecular polarisation relaxation, i.e. relaxation of the positive core-hole, remaining after photoemission, by polarisation of the surrounding medium and it accounts for the marked increase in linewidth between gaseous and condensed species (Beamson & Briggs, 1992b). Finally, if sufficient care is not paid to effective charge neutralisation when using a monochromator the result will be some additional broadening, probably accompanied by asymmetry (Beamson *et al.*, 1990).

The overall lineshape, then, is a complex convolution of mainly mixed Gauss–Lorentzian character with the possibility of weak asymmetry (tailing to high binding energy which is only likely to be manifested at high resolution). The Gauss–Lorentzian ratio will depend on the relative importance of the instrumental contributions.

It is therefore prudent to ascertain the Gauss–Lorentzian ratio which results from the instrumental parameters routinely used for polymer core level studies; these will generally be fixed at some optimal values which take into consideration the trade-off between minimum fwhm (i.e. highest energy resolution) and count-rate (which, in turn, involves signal:noise, acquisition time and radiation damage potential). A suitable polymer is one in which the chosen core level represents a single chemical state, or one which is well resolved from other

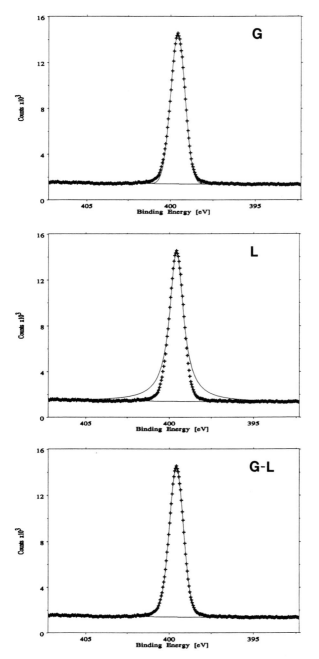

Figure 3.3 Experimental N1s peak from poly(arylonitrile) fitted with pure Gaussian (G), pure Lorentzian (L) and mixed G–L (83%G) lineshapes.

components. For reasons discussed below this should not be a CH_x functionality in the case of C1s. As an example Fig. 3.3 shows the N1s peak from poly(acrylonitrile), $(CH_2-CHCN)_n$, fitted with pure Gaussian, pure Lorentzian and mixed Gauss–Lorentzian (83% Gaussian) lineshapes. Several such measurements will define the appropriate starting lineshape and a reasonable

Gauss–Lorentzian ratio range (if this is an allowed variable) for the components in any attempted curve fitting process (Section 3.4.2).

All of the data discussed in this chapter are derived from a single instrument (the prototype Scienta ESCA 300) which set new standards for the highest resolution spectra from polymers. A large database of Scienta spectra from over 100 pure homopolymers has been published together with detailed descriptions of the instrument performance, standard operating conditions and data interpretation procedures. (Beamson & Briggs, 1992a). These details do not need to be repeated here. Since this instrument could operate with both monochromated and non-monochromated Al$K\alpha$ radiation it is interesting to compare spectra obtained from the same polymer under otherwise identical operating conditions (except for the use of an electron flood gun with the monochromated source). This is done for PET in Fig. 3.4.

3.4.1.1 Vibrational excitation

Vibrational fine structure in the C1s spectrum of gas phase methane was observed over 20 years ago (Gelius *et al.*, 1974) and the effect has been seen in other gaseous molecules since then. The first observation of the effect in the solid state came with the application of the high resolution instrument described above (Beamson *et al.*, 1991).

Fig. 3.5 compares the C1s spectra of solid hexatriacontane (n-$C_{36}H_{74}$) and polystyrene ($-CH_2-CH(C_6H_5)-)_n$ run as very thin films on silicon and gaseous methane (CH_4). In the methane spectrum four components are resolved with the relative intensities shown and a constant separation of 0.39 eV. Because of relaxation effects, core ionised CH_4^+ has an equilibrium C$-$H bond length 0.05 Å shorter than in CH_4 and a narrower potential curve. It therefore follows from the Franck–Condon principle that C1s photoemission will result in the population of vibrational states above the ground state in the ion, i.e. photoemission is accompanied by vibrational excitation (Gelius, 1974). Thus the four components of the C1s spectrum have intensities which reflect the relative populations of the vibrational levels $v=0-3$, of increasing binding energy, respectively, and the separation is the symmetric C$-$H stretch energy in CH_4^+. If the hexatriacontane spectrum is curve fitted (see below) with four components of equal fwhm and a fixed separation of 0.39 eV the relative intensities are almost exactly those seen in methane. Therefore the asymmetry of the C1s line is most probably due to the same vibrational fine structure, although there may be weaker contributions from lower frequency vibrations such as CH_2 bend (0.19 eV) and C$-$C stretch (0.15 eV). Similar fine structure needs to be invoked to fit the C1s spectra of poly(ethylene), poly(propylene) and poly(lauryl metharylate) accurately even with the somewhat lower resolution pertaining to these fully insulating samples (Beamson *et al.*, 1991; Beamson and Briggs, 1992a). The poly(styrene) spectrum shown in Fig. 3.5, whilst having (fortuitously) the same fwhm as that from

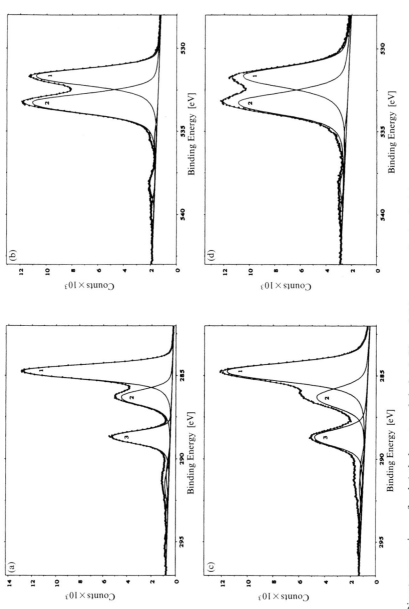

Figure 3.4 Direct comparison of poly(ethylene terephthalate) C1s and O1s spectra obtained with monochromated ((a),(b)) and non-monochromated ((c),(d)) Al $K\alpha$ radiation with otherwise identical instrumental parameters. Note the $\sim 50\%$ increase in the background intensity in the latter case (Beamson & Briggs, 1992a).

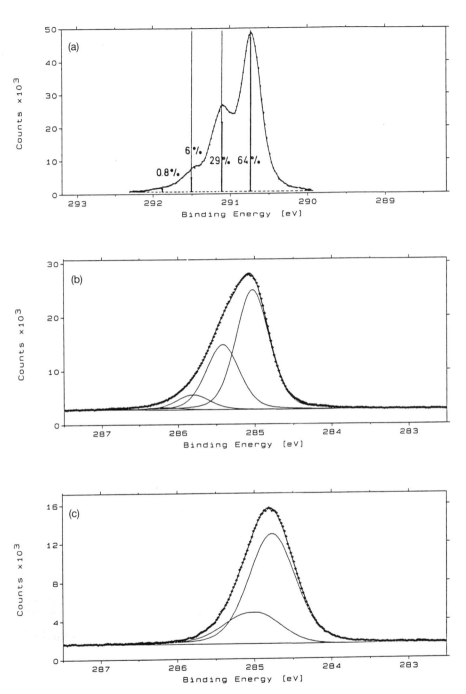

Figure 3.5 High resolution C1s spectra of hydrocarbons: (a) CH_4 gas phase, after Gelius *et al.* (1984); (b) thin film of hexatriacontane (n-$C_{36}H_{74}$) on silicon; (c) thin film of poly(styrene) on silicon (Beamson & Briggs, 1992a).

hexatriacontane, is much more symmetrical. It can be fitted with two envelopes with approximate relative intensities of 6:2; the low binding energy component (due to the ring carbons) is symmetric whilst the higher binding energy component (due to the backbone carbons) is asymmetric in the manner just described. In the case of the ring C−H bonds, relaxation effects largely involve the π-system and result in well-defined shake-up structure (Section 3.5) rather than the vibrational broadening. Finally, it should be noted that for more complex polymers, giving C1s components which represent aliphatic CH_x containing functions (e.g. $-CH_2OCH_3$) a single asymmetric component can be used in curve fitting to a good approximation (this is *not* the case for polymers mentioned above). Examples of such situations appear later in this chapter.

The corollary is that aliphatic carbon atoms not attached to hydrogen will not be affected by C−H vibrational excitation and should therefore give rise to narrower, more symmetrical C1s peaks. This is clearly demonstrated by the PET spectrum (Fig. 3.4) in which the carboxyl component (3) is significantly narrower (by >0.1 eV) than the other two. Similarly, quaternary carbons will also be unaffected. These effects are further evidenced in Section 3.4.3.

It seems likely that vibrational fine structure effects are also partly responsible for variable O1s peak widths, with the fwhm for this peak in polysiloxanes (involving Si−O−Si) significantly narrower than in carbon polymers (Beamson & Briggs, 1992a). It is noticeable that the two components of the O1s doublet from carboxyl groups, −COOR, are of different width (Fig. 3.4 and more obviously in Fig. 3.6). In the latter case the relative widths are invariant to the separation, which is discussed further in Section 3.4.4. In the equivalent cores approximation, core ionised oxygen is electronically equivalent to fluorine ($O^+ \equiv F$). One of the bonds in C−O−R is dissociative in the final state following O1s photoemission, whereas C=O is still a bound state. Using the same arguments involving the Franck–Condon principle as above, the latter will lead to a smaller peak width.

3.4.2   Curve fitting

It is appropriate at this point to discuss briefly curve fitting techniques, without which a detailed interpretation of core level spectra is impossible. This approach should not be confused with deconvolution which aims to remove instrumental broadening from the measured spectrum in order to enhance the energy resolution (to date with very limited success). XPS data analysis, particularly curve fitting and the mathematical basis of the techniques used, has been discussed in detail by Sherwood (1990).

Prior to curve fitting an appropriate background function (Section 2.2.3.1) will have been introduced (or used for background removal) and low signal:noise data may have been smoothed. Clues as to the likely position of

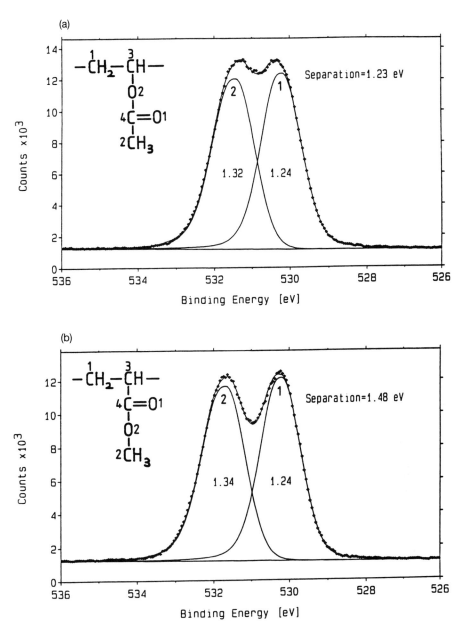

Figure 3.6 High resolution O 1s spectra from thin films on silicon of (a) poly(vinyl acetate) and (b) poly(methyl acrylate) (Briggs & Beamson, 1993).

component maxima can often be gained by taking the second derivative of the original spectrum. In curve fitting itself a series of components is generated and placed within the measured envelope at appropriate binding energy positions. Each component will have an initial lineshape determined on the basis of criteria discussed above and an intensity (area) which seems reasonable.

The software calculates the sum of the components and from comparison with the experimental data obtains a measure of the 'goodness of fit' (usually chi-squared by the least squares method). By iteration, with variation of allowed parameters describing the peak position, width, intensity etc, the fit is refined until convergence to a minimum χ^2 results. It cannot be overemphasised that this process cannot produce a unique solution and that a great deal of operator input is usually required to ensure that the solution is spectroscopically and chemically reasonable. Good software will allow: the generation of line-shapes with variable Gauss–Lorentzian ratio and asymmetry; the variation of these parameters together with peak fwhm, position (of maximum or centroid) and intensity as well as background position during iteration; the ability to 'freeze' certain parameters whilst others vary and to 'lock' certain parameters into some predetermined relative values (e.g. energy separation or intensity ratio). In this way the fitting process can be 'guided' towards a solution which incorporates the maximum amount of operator understanding about the system.

The fitting process used for the examples shown in this chapter has been fully described (Briggs & Beamson, 1992a).

3.4.3   The chemical shift effect

The fact that the same atom in different chemical environments can give rise to discrete components in its core level spectrum gave an early, and unexpected, boost to the development of XPS. By analogy with nmr spectroscopy this was dubbed a chemical shift effect and, to the extent that electron density on the atom is an important parameter in determining the chemical shift in both spectroscopies, the analogy has some validity. However, in the case of XPS both 'initial state' and 'final state' effects are involved; it is important to recognise that the core level spectrum is actually that of the ionised system.

If, for simplicity, we consider photoemission from a core orbital of a gas phase molecule, where energies are absolute and referenced to the vacuum level, then by conservation of energy:

$$E_I - h\nu = E_F - E_K \qquad (3.1)$$

where E_I and E_F are the initial (before photoemission) and final state energies, respectively, and the outgoing electron has kinetic energy, E_K. Since

$$\text{binding energy, } E_B = h\nu - E_K = E_F - E_I \qquad (3.2)$$

it follows that the measured binding energy is the difference between final and initial state energies, and that chemical shifts must involve consideration of both of these. The physical basis of initial state effects can be discussed in terms of the charge potential model (Clark, 1973).

$$E_i = E_i^\circ + kq_i + V_i \quad \text{where} \quad V_i = \sum_{i \neq j} \frac{q_i}{r_{ij}} \tag{3.3}$$

where E_i is the binding energy of a particular core level of atom i, E_i° is an energy reference, k is an adjustable parameter, q_i is the (partial) charge on atom i and V_i sums the potential at atom i due to 'point charges' on surrounding atoms j. In a molecular solid these are basically the atoms bonded to atom i and V_i is generally opposite in sign to kq_i (in an ionic solid V_i is summed over the entire lattice and is closely related to the Madelung energy of the solid, hence it is often referred to as a Madelung potential). The model considers the atom to be a hollow sphere on the surface of which the valence charge q_i resides. The electrical potential inside the sphere is the same at all points and equal to q_i/r_v, where r_v is the average valence orbital radius. Redistribution of the valence electrons (new compound formation) leads to a change in the surface charge, Δq_i, and a change of potential, $\Delta q_i/r_v$, inside the sphere. Thus all core levels will experience a change in binding energy by this amount (a good approximation in practise). From equation (3.3) the chemical shift in binding energy for atom i in two different environments is

$$\Delta E_i = k\Delta q_i + \Delta V_i \tag{3.4}$$

For molecular solids in general, and polymers in particular, ΔV_i is relatively small and electronegativity effects will dominate Δq_i (an increase in positive charge leading to an increased binding energy). Hence initial state effects can largely be accounted for by consideration of the relative electronegativities of atoms or groups bonded to atom i.

In order to consider final state effects it is first necessary to discuss two limiting approximations to the photoionisation/photoemission event, namely the *adiabatic* and *sudden* approximations. In the case of the former, the total process takes place in such a way that equilibrium is maintained and the final state is the ground state of the ion. In the latter case, the process takes place very rapidly such that the final state may be electronically excited, giving rise to discrete features on the low kinetic energy side of the primary peak. These 'shake-up' and 'shake-off' processes are discussed in more detail in Section 3.5.

The adiabatic approximation is useful for understanding final state contributions to chemical shifts. The formation of a core-hole through photoemission decreases the screening of the nuclear charge, and the remaining electrons 'relax' in order to minimise the total energy of the system. The relaxation energy (typically 10–20 eV) is added to the kinetic energy of the photoelectron and it can be divided into two parts: that due to intra-atomic relaxation (of electrons localised

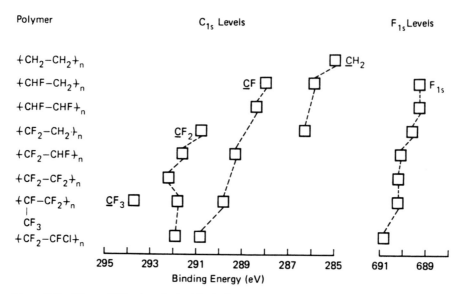

Figure 3.7 C1s and F1s chemical shifts in fluoropolymers (Dilks (1981) – from original data in Clark *et al.*, 1973).

on atom *i*) and to extra-atomic relaxation (of electrons on atoms *j* involved in bonding orbitals). The magnitude of the latter is dependent on the polarisability of the system in response to formation of the core-hole and this is obviously structure sensitive. If the two contributions to the relaxation energy are E_i^{IA} and E_i^{EA} respectively (both of which increase the photoelectron kinetic energy, or decrease the binding energy), the contributions to the chemical shift are $-\Delta E_i^{\text{IA}}$ (approximately zero) and $-\Delta E_i^{\text{EA}}$. Adding the final state effects to those of the initial state (equation (3.4)) then gives a binding energy shift:

$$\Delta E_i = k\Delta q_i + \Delta V_i - \Delta E_i^{\text{EA}} \tag{3.5}$$

Since the charge potential and extra-atomic relaxation terms are of similar magnitude (typically in the range 0–10eV), the origin of binding energy variations between different functionalities can often defy a simple explanation.

3.4.4 The C1s level

The earliest studies of C1s chemical shifts were carried out on fluoropolymers since fluorine, being the most electronegative element, induces the largest shifts. A compilation of these data (Dilks, 1981) is reproduced in Fig. 3.7. This shows clearly that fluorine exerts both a primary substituent effect (on the carbon atom to which it is directly bound) and a secondary substituent effect (on the carbon atom in the chain one removed from the primary carbon). The latter is also

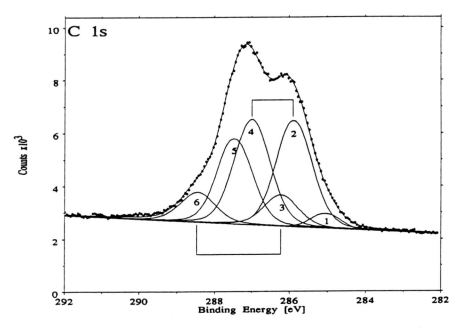

Figure 3.8 Curve fit to the C 1s spectrum of chlorinated poly(propylene): see text for component details (Beamson & Briggs, 1992a).

referred to as a β shift. For a single substitution of hydrogen by fluorine the primary shift is ~2.9 eV and the secondary shift ~0.7 eV. For this closely related series of polymers the C 1s binding energies can be well estimated for increasing substitutions simply by adding together the appropriate multiples of primary and secondary shifts. This suggests that relaxation effects are fairly constant, so that the chemical shift is dominated by σ-inductive effects. Similarly, for chlorine substitution the primary shift is ~2.0 eV and the secondary shift ~0.5 eV reflecting the lower electronegativity of chlorine relative to fluorine; for bromine the primary shift has reduced to ~0.7 eV.

The internal consistency of these concepts in analysing a complex surface is illustrated by chlorinated poly(propylene) (Beamson & Briggs, 1992a). Neglecting chlorination of the methyl group (from other evidence) the functionalities which may be present are:

$$
\begin{array}{cccccc}
\overset{1}{} & \overset{1}{} & \overset{2}{} & \overset{4}{} & \overset{5}{}\quad\overset{5}{} & \overset{3}{}\quad\overset{6}{}
\end{array}
$$

$$
CH_3 - CH - , \quad -CR_2 - \underset{|}{CR} - , \quad -\underset{|}{CR} - \underset{|}{CR} - , \quad CR_2 - CCl_2
$$

$$
Cl Cl \quad Cl
$$

where R = H or CH$_3$.

The C 1s envelope (Fig. 3.8) can then be fitted with components at appropriate starting positions based on the above shift values, locked into 1:1 relative intensity ratios where the stoichiometry demands (peaks 2/4 and 3/6 representing the structures shown above). This leads, on iteration, to an excellent fit with the components parameterised as shown in Table 3.2. From these data the chlorine concentration can be calculated to be 41 at%, which compares well with the 39 at% calculated directly from the C1s and Cl2p intensities (as in Section 2.2.3.3). As in all such XPS determinations of surface composition the hydrogen content is not included.

For carbon bound to oxygen the number of functionalities is greatly increased. As a very crude approximation, the primary shift is ~1.4eV per C—O single bond (i.e. double this for C=O etc). From a detailed study of 43 polymers containing only oxygen functionalities Briggs & Beamson (1992) compiled the summary of primary shifts given in Table 3.3. The results also showed that secondary shifts involving oxygen are significant and these are summarised in Table 3.4. The last two entries reflect the difference in measured β shifts in acrylate and methacrylate polymers. An interesting spectrum from a polymer containing several oxygen functionalities and exhibiting a range of primary and/or secondary shifts is shown in Fig. 3.9. Parameters relating to the curve fit are given in Table 3.5. The result in terms of component binding energies could have been fairly accurately predicted from the summary data in Tables 3.3 and 3.4 and the relative fwhm values accord with the observations made in Section 3.4.1.1. The relative intensity data are discussed in Section 3.5.

Many nitrogen-containing functions also contain oxygen, and a combined summary of C1s shifts is presented in Table 3.6. Note that the —C≡N group behaves as a pseudo-halogen.

The shifts involved in most other functional groups are fairly small. It is important, however, to comment on the situation for functions involving different hybridisations of carbon when bound to itself and/or hydrogen. It

Table 3.2 *Parameters for the component C1s peaks used in Fig. 3.8; m is the Gauss–Lorentzian mixing ratio (where 1.0 is pure Gausian) and A is an asymmetry parameter defined in Beamson & Briggs (1992a)*

	C1s component					
	1	2	3	4	5	6
binding energy (eV)	285.00	285.86	286.22	286.99	287.44	288.44
fwhm (eV)	0.92	1.14	1.09	1.14	1.12	1.10
area (%)	3	28	8	28	23	8
A	0.07	0.01	0.00	0.00	0.03	0.00
m	1.00	0.85	0.79	0.84	0.82	0.71

used to be believed that all such functions would give rise to the same (unshifted) C 1s binding energy=285.00 eV. However, it now seems certain that unsubstituted aromatic carbon atoms have a slightly lower binding energy (average shift~–0.3 eV), as seen above for poly(styrene). Moreover, the high resolution C 1s peaks from unsaturated aliphatic hydrocarbon

Table 3.3 *Primary C 1s chemical shifts (eV) for oxygen functions, relative to saturated hydrocarbon (C 1s=285.00 eV) (Beamson & Briggs, 1992a)*

Functional group	Chemical shift			Number of examples
	Min.	Max.	Mean	
C—O—C	1.13	1.75	1.45	18
C—OH	1.47	1.73	1.55	5
*C—O—C[a] (‖O)	1.12	1.98	1.64	21
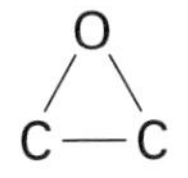 O / C—C (epoxide)	—	—	2.02	1
C=O[b]	2.81	2.97	2.90	3
O—C—O	2.83	3.06	2.93	5
O—C—*C[c] (‖O)	3.64	4.23	3.99	21
HO—C— (‖O)	4.18	4.33	4.26	2
O—C—O (│O)	4.30	4.34	4.32	2
—C—O—C— (‖O ‖O)	4.36	4.46	4.41	3
—O—C—O— (‖O)	5.35	5.44	5.40	2

[a]Neglecting aromatic carboxylic esters, mean of 18 is 1.72, min. 1.48.

[b]PEEK significantly lower: shift=2.10 (binding energy) referenced to aromatic CH C 1s=284.70 eV).

[c]Neglecting aromatic carboxylic esters, mean of 18 is 4.05, min. 3.84.

polymers (e.g. poly(isoprene)) are systematically asymmetric, indicating a similar negative shift for the carbon atoms involved in C=C units (Beamson & Briggs, 1992a). These, and other miscellaneous C1s shifts are summarised in Table 3.7.

There are some particular features to mention relating to C1s shifts in aromatic polymers. There may be significant differences between the binding energies of an aromatic carbon bound to a heteroatom and its aliphatic analogue. Usually the shifts will be lower in the former case because of delocalisation effects. The C=O C1s binding energy in PEEK (poly(etheretherketone)) is ~287.3 eV, about 0.7 eV lower than expected, presumably because the carbonyl group is conjugated to the attached ring. In polyimides, in which N-phenylimide groups are fused to a benzene ring:

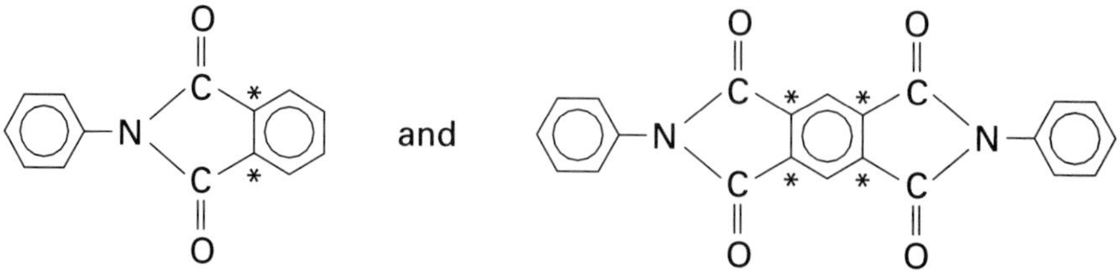

the ring carbons (*) at the point of attachment are shifted by ~0.5 eV by one imide group and by ~1.0 eV by two groups (relative to unshifted phenyl C1s at ~284.7 eV)).

Table 3.4 *Secondary C1s chemical shifts (eV) relative to saturated hydrocarbon (C1s=285.00 eV) (Beamson & Briggs, 1992a)*

Functional group	Chemical shift	Number of examples
O—C—*C	~0.2	6
F—C—*C	~0.4	11
Cl—C—*C	~0.5	9
—C—*C (C=O)	~0.4	4
O—C—*C (C=O)	~0.4	7
O—C—*C—CH₃ (C=O)	~0.7	8

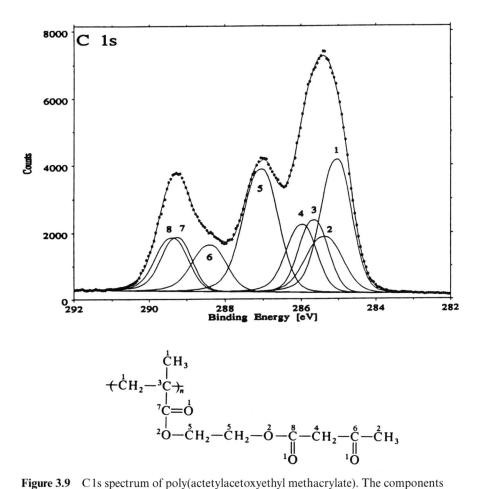

Figure 3.9 C 1s spectrum of poly(actetylacetoxyethyl methacrylate). The components refer to the carbon atoms numbered in the repeat unit formula, the relative ordering of 7/8 being uncertain (Briggs & Beamson, 1992a).

3.4.5 The O 1s level

In terms of its usefulness for directly identifying oxygen functionalities, the O 1s level is disappointing. Compared with the C 1s level, the inherent peak width is significantly greater, in most cases, and the dynamic range more restricted. This can be appreciated from Table 3.8 where the range of mean values covers only 2.7 eV (note also the difference in binding energy between aliphatic and aromatic situations). Most other oxygen functionalities, involving other heteroatoms, also fall within this same range. The only exceptions are nitrate ester, the 1:2 O 1s doublet components of $R-O-NO_2$ appearing at $\sim$533.9 and 534.7 eV, respectively, and perfluoroether, with an O 1s BE of $\sim$535.7 eV. Results from over 70 oxygen-containing polymers have been discussed by Briggs & Beamson (1993).

Table 3.5 *Parameters for the component C1s peaks used in Fig. 3.9; m is the Gaussian–Lorentzian mixing ratio (where 1.0 is pure Gaussian) and A is an asymmetry parameter defined in Beamson & Briggs (1992a)*

	C1s component							
	1	2	3	4	5	6	7	8
binding energy	285.00	285.33	285.63	285.94	287.01	288.36	289.22	289.35
fwhm	0.99	1.18	0.93	0.96	1.08	1.06	0.95	1.02
area (%)	21	11	11	11	21	8	8	9
A	0.10	0.08	0.07	0.10	0.10	0.12	0.12	0.11
m	0.91	0.92	0.97	0.92	0.92	0.92	0.94	1.00

It is interesting to note the unpredictability of the O1s binding energy in C=O units, which is believed to be due to final state (relaxation) effects involving the system outweighing initial state effects (Section 3.4.3). This is further demonstrated by the binding energy difference between the singly and doubly bonded O atoms in the O=C−O unit in different environments. Fig. 3.10 shows the O1s doublet for poly(vinylacetate), poly(caprolactone), poly(methyl acrylate) and poly(urethane) in which the splitting is 1.23, 1.30, 1.48 and 1.73 eV, respectively.

3.4.6 The N1s level

Most nitrogen functions in which nitrogen is bound to carbon give very similar N1s binding energies in the range 399–400 eV. This includes amine, amide,

Table 3.6 *Primary C1s chemical shifts (eV) for nitrogen functions, relative to saturated hydrocarbon (C1s=285.00 eV) (Beamson & Briggs, 1992a)*

Functional group	Chemical shift			Number of examples
	Min.	Max.	Mean	
$C-NO_2$ [a]	–	–	0.76	1
$C-N\langle$	0.56	1.41	0.94	9
$C-\overset{+}{N}\langle$	0.99	1.22	1.11	2
$*C-C\equiv N$	1.35	1.46	1.41	2
$-C\equiv N$	1.73	1.74	1.74	2
$C-ONO_2$	–	–	2.62	1
$N-C-O$	–	–	2.78	1
$N-C=O$	2.97	3.59	3.11	6
$\begin{matrix} C-N-C \\ \| \quad\quad \| \\ O \quad\quad O \end{matrix}$	3.49	3.61	3.55	2
$\begin{matrix} N-C-N \\ \| \\ O \end{matrix}$	–	–	3.84	1
$\begin{matrix} N-C-O \\ \| \\ O \end{matrix}$	–	–	4.60	1

[a] C in phenyl ring.

nitrile, urea and nitrogen in aromatic rings. The binding energies of carbamate and imide are usually ~400.5 eV. For polyphosphazenes, involving the P=N–P linkage, the N1s binding energy is significantly lower at ~397.9 eV. Quaternary nitrogen gives binding energies which are higher, as expected from the localised positive charge, but the alkyl substituents and counter ion can lead to a range of binding energies of ~401.5–402.5 eV. Much higher binding energies require higher oxidation states, thus nitro $-NO_2$ and nitrate $-ONO_2$ give N1s binding energies of ~405.5 and 408.2 eV, respectively (Beamson & Briggs, 1992a).

Table 3.7 *Primary C1s chemical shifts (eV) for miscellaneous and halogen functions, relative to saturated hydrocarbon (C1s=285.00 eV) (Beamson & Briggs, 1992a)*

Functional group	Chemical shift			Number of examples
	Min.	Max.	Mean	
C=C	−0.24	−0.31	−0.27	4
(phenyl)–X [a]	−0.20	−0.56	−0.34	20
C−Si	−0.61	−0.78	−0.67	3
C−S	0.21[b]	0.52	0.37	2
C−SO$_2$	0.31[b]	0.64	0.38	2
C−SO$_3^-$ [b]	–	–	0.16	1
C−Br [b]	–	–	0.74	1
(phenyl)*−Cl	0.99	1.07	1.02	3
C−Cl	2.00	2.03	2.02	2
−CCl$_2$	–	–	3.56	1
C−F	–	–	2.91	1
−CF$_2$	–	–	5.90	1
−CF$_3$	7.65	7.72	7.69	2

[a] Average of C atoms not directly bonded to substitutent X.

[b] In this example bonded C atom is in phenyl ring.

Table 3.8 *O1s binding energies in polymers containing C, H and O, relative to saturated hydrocarbon (C1s=285.00 eV) (Beamson & Briggs, 1992a)*

Functional group		Binding energy			Number of examples
		Min.	Max.	Mean	
C—O—C (aliphatic)		532.47	532.83	532.64	8
(aromatic)[a]		532.98	533.45	533.25	3
C—OH (aliphatic)		532.74	533.09	532.89	4
(aromatic)[b]				533.64	1
C—C (epoxide)				533.13	1
O—C—O		532.94	533.51	533.15	5
O—C—O with O		532.95	533.02	532.99	2
C=O (aliphatic)		532.30	532.37	532.33	3
(aromatic)[c]				531.25	1
Aliphatic O^2—C—C with O^1	1	531.96	532.43	532.21	20
	2	533.24	533.86	533.59	
Aromatic O^2—C—Ar with O^1	1	531.62	531.70	531.65	3
	2	533.06	533.22	533.14	
—C—O^2—C— with O^1 O^1	1	532.52	532.81	532.64	3
	2	533.86	534.02	533.91	
O^2—C—O^2 with O^1	1	532.33	532.44	532.38	2
	2	533.88	533.97	533.93	

[a]One or both carbon atoms in phenyl ring (PMS (poly(4-methoxystyrene)), PODMPO (poly(2,6-dimethyl-1,4-phenylene oxide)), PEEK (poly(ether ether ketone))).

[b]Carbon atom in phenyl ring (PHS (poly(hydroxy styrene))).

[c]Carbonyl group bonded to phenyl ring (PEEK).

For PEEK and PET binding energy are referenced to aromatic CH C1s=284.70eV.

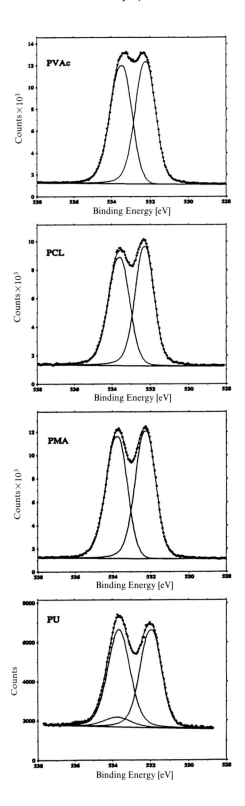

Figure 3.10 O 1s doublets for poly(vinyl acetate) (PVAc), poly(caprolactone) (PCL), poly(methyl acrylate) (PMA) and a poly(urethane) (PU) not containing any other oxygen functionality. The small third component in the PU spectrum is due to $\pi \rightarrow \pi^*$ shake-up (Briggs & Beamson, 1993).

3.4.7 Other core levels

The only other core level encountered in polymer XPS which exhibits shifts due variable oxidation state is S 2p. Despite the fact that this is a 2:1 spin–orbit doublet (separation 1.9 eV) it is studied in preference to the S 2s singlet because of its higher cross-section. The S $2p_{3/2}$ binding energy covers a reasonable range: ~163.5 eV in sulphide (RSR), ~167.6 eV in sulphone (RSO_2R) and 168–169 eV in sulphonate/sulphonic acid (RSO_3^-/RSO_3H).

It can be seen from Fig. 3.7 that the F 1s binding energy is dependent on the degree of fluorination of a polymer. In polymers with a range of fluorine environments this is manifested as a broadened peak. Thus, for instance, in poly(vinyl fluoride), $(CH_2CHF)_n$, the F 1s binding energy is 686.94 eV and the fwhm is 1.45 eV, whereas in Viton A, which has the structure.

$$\text{---}\!\!\underset{\substack{|\\CF_3}}{(CF-CF_2)_x}\!\!(CF_2-CH_2)_y\text{---}$$

the F 1s binding energy is 688.80 eV and the fwhm is 2.01 eV (under the same instrumental conditions) (Beamson & Briggs, 1992a). If the F 1s binding energy in fluoropolymers is taken to be typically 689 eV then fluoride ion, F^-, appears at a binding energy some 4 eV lower (depending on the counter ion and local environment) and as a distinctly separate peak from C–F.

In chlorine-containing polymers the situation is less complicated. The Cl $2p_{3/2}$ level is used (as in the case of sulphur discussed above), the spin–orbit doublet separation being 1.6 eV. In covalent bonding the Cl 2p binding energy is almost constant at 200.6±0.2 eV, whilst chloride ion, Cl^-, has a binding energy some 3–4 eV lower (again, this is very dependent on the local environment).

3.5 Shake-up satellites

Final state relaxation effects were introduced in Section 3.4.3. In the sudden approximation, the relaxation of the valence electrons in response to photoemission of a core-electron (which is almost completely screening as far as the valence electrons are concerned) is such that the final state may involve an electron in an excited bound state (a process referred to as 'shake-up') or in an unbound state above the vacuum level (a process referred to as 'shake-off'). The former leads to discrete peaks (satellites) on the low kinetic energy side of the main core level, the energy separation being that of the valence electron transition involved. The latter leads to a step increase in intensity at some point on the lower kinetic energy side of the main peak because the unbound states involved form a continuum. The energies involved in these valence electron transitions are generally

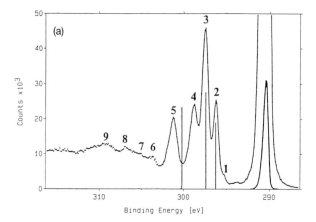

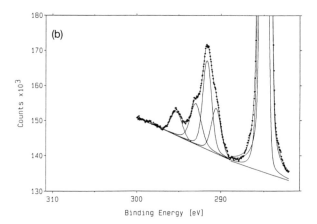

Figure 3.11 C1s shake-up and elastic scattering spectra for: (a) gas phase benzene at low pressure, after Nordfors *et al.* (1988) and (b) an ~3 nm film of polystyrene on silicon, inelastic scattering minimised by using 90° detection (relative to the surface) (Beamson & Briggs, 1992b).

<15 eV. Unfortunately, in the same region of the core level spectrum there is a broad feature due to inelastic scattering (see Section 2.2.2), which for organic materials peaks at ~20 eV below the main peak. This steep rise in the local background obscures shake-off features entirely and limits the observation of shake-up satellites to those of relatively high intensity and low energy (and hence a small separation from the main peak).

The conditions for observing shake-up satellites are best met by aromatic polymers (either backbone or side-chain rings), of which a good prototype example is poly(styrene). Fig. 3.11 compares the high resolution XPS spectra in the C1s region for low pressure, gas phase benzene and poly(styrene). Apart from the solid-state broadening (see Section 3.4.1) the two spectra are very similar, in terms of the separation of the shake-up components from the main peak and their relative intensities. This is because the transitions involved are of the $\pi \rightarrow \pi^*$ variety; in poly(styrene) the pendant group is essentially independent, the ring π orbitals being relatively unperturbed by the attached backbone. The overall shake-up intensity in polystyrene is almost 10% that of the main peak and the separation of the most intense component is 6.8 eV. It is worth noting that this component is enhanced in intensity by ~50% in the spectrum from a thick film because it coincides with a discrete energy loss feature (the transition from the highest occupied molecular orbital to the lowest unoccupied molecular orbital (HOMO–LUMO) in the neutral molecule excited by C1s electrons passing through the polymer film (Beamson & Briggs, 1992b). The analogous effect is seen in the gas phase benzene spectrum when the gas pressure is increased. $\pi \to \pi^*$ shake-up can also be observed in the C1s spectra from unsaturated hydrocarbon polymers such as poly(isoprene), but with an intensity of only ~4% relative to the main peak.

Discrete shake-up satellites can also be observed, albeit with more difficulty, for polymers containing the C=O group; again these are $\pi \to \pi^*$ transitions. Fig. 3.12 compares the C1s and O1s spectra from poly(vinyl methyl ketone) (PVMK) and gas phase acetone:

$$\begin{array}{cc}
\text{---}(\text{CH}_2\text{---}\text{CH})_n\text{---} & \text{CH}_3 \\
\qquad\quad | & | \\
\qquad\quad \text{C}=\text{O} & \text{C}=\text{O} \\
\qquad\quad | & | \\
\qquad\quad \text{CH}_3 & \text{CH}_3 \\
\\
\text{PVMK} & \text{acetone}
\end{array}$$

The shake-up structures in the O1s spectrum of PVMK are in the same position as for acetone, and although the overall intensity of shake-up is comparable the relative intensity distribution is significantly different (this is not surprising since acetone is a poorer model for PVMK than benzene is for poly(styrene)). However, in the C1s spectrum of PVMK the shake-up is almost invisible. This is probably because only the carbonyl carbon contributes to the shake-up structures whilst all four carbons in the repeat unit contribute to the inelastic scattering background on which the shake-up satellites sit. In the case of the O1s spectrum the single oxygen contributes to both shake-up and background features. This is confirmed by studies of PMMA where the less distinct O1s shake-up involving the C=O unit sits on an inelastic background resulting from the two oxygen atoms in the carboxyl group (Beamson & Briggs, 1992b).

The gas phase spectra of several small molecules, such as benzene and acetone shown above, have been compared with theoretical calculations to identify the transitions involved. These calculations represent the final state as a linear superposition of the (N-1) electron states in the ion resulting from the N electron states of the neutral molecule and compute the coefficients which leads to minimisation of the total energy. From these the relative intensities of the various shake-up and shake-off states can be estimated. This procedure is known as a final state configuration interaction (FSCI) calculation and FSCI is an alternative description for shake-up and shake-off. Clearly these calculations and the gas phase model spectra are very helpful in the interpretation of polymer shake-up spectra (for more details see Nordfors et al. (1988) and Keane et al. (1991)).

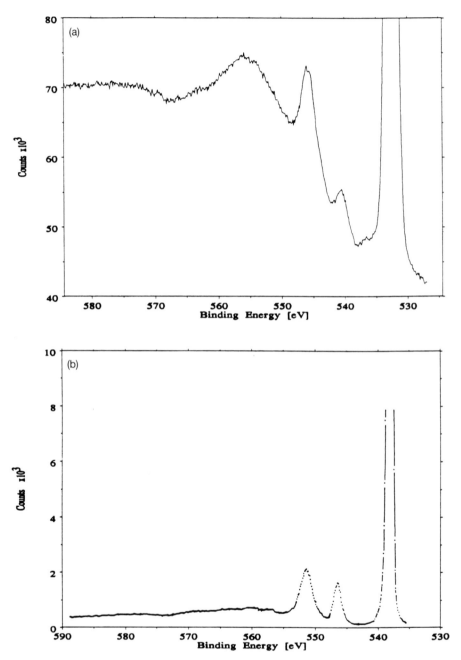

Figure 3.12 Shake-up/inelastic scattering spectra for PVMK (thick film on glass) and gas phase acetone: (a) PVMK O 1s; (b) acetone O 1s; (c) PVMK C 1s; (d) acetone C 1s (acetone spectra after Keane *et al.* (1991)) (Beamson & Briggs, 1992b).

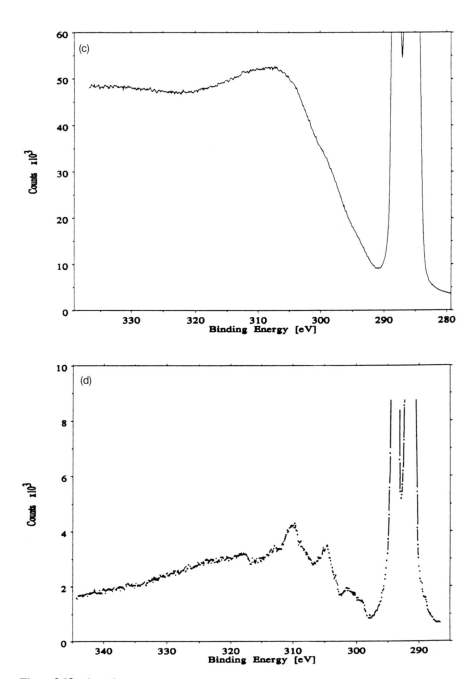

Figure 3.12   (*cont.*)

If molecular orbitals involving more than one type of atom participate in a shake-up transition it follows that the core levels of all such atoms will be accompanied by the same satellite (appropriately weighted in intensity). This spectroscopic correlation generally indicates direct bonding of the atoms concerned. The spectra of ring-substituted poly(styrenes) show this clearly: for instance the three poly(chlorostyrenes) exhibit subtly different shake-up spectra (in terms of band profile and relative intensity to the main peak) in both the C 1s region (8–10% shake-up intensity) *and* the Cl 2p region (4–5% shake-up intensity) (Beamson & Briggs, 1992a).

An important consequence of shake-up (and shake-off) is the loss of intensity from the main peak and its effect on quantification. As we have seen, even under optimum conditions of high signal:noise (helped by use of monochromatic radiation) non-aromatic $\pi \rightarrow \pi^*$ shake-up may be virtually impossible to detect directly, and this is certainly true of shake-off. For example, the data in Table 3.5 reveal the non-stoichiometry in the C 1s components representing C=O/O–C=O groups and a study of many polymers containing these functions suggested on average a 13% underestimation (Briggs & Beamson, 1992). In cases such as PVMK at least half of this can be associated directly with shake-up. Similarly in the O 1s spectra of carboxyl containing polymers the C=O component is less intense than the C–O component (see Fig. 3.6 in which the peak heights are the same but the C=O component is significantly narrower). This complication can become extreme in conjugated aromatic polymers in which some $\pi \rightarrow \pi^*$ transitions have very low energy, such that they overlap with the main core level envelope (especially if high binding energy components are present), and relatively high intensity. Examples are conducting polymers generally, polyimides (see Fig. 3.13), polyketones and polysulphones.

3.6   Valence band spectra

Photoelectron emission from the molecular orbitals involved in chemical bonding gives rise to the valence band spectrum. In principle this should be more sensitive to molecular structure than the core lines since the latter only indirectly reflect changes in the valence electron distribution (Section 3.4.3). For example Fig. 3.14 shows the valence band spectra for poly(ethylene), poly(propylene) and poly(but-l-ene), all of which give identical C 1s spectra even under the highest resolution conditions. In fact, detailed studies of valence band spectra (using monochromated X-rays) by Pireaux *et al.* (1981) showed early on in polymer XPS that they were sensitive to various types of isomerism (structural-, linkage- and stereo-) as well as to tacticity and geometrical conformation.

The valence band of a polymer is typically 20–100 times weaker (based on the most intense feature) than the principal core lines. This implies rather long acquisition times with consequent concerns about surface potential stability

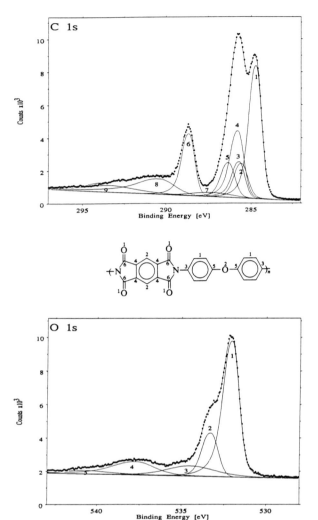

Figure 3.13 Shake-up satellites in the C 1s (components 7–9) and O 1s (components 3–5) spectra of the poly(ether imide) Kapton HN.

(charging) and sample damage. Therefore until the advent of much enhanced instrumental sensitivity, valence bands were largely neglected. The advantages of using monochromated radiation in their study are threefold: (1) features in the valence band can be as narrow as the core lines and these are better revealed with high energy resolution; (2) the spectrum can cover up to 40 eV so that X-ray satellites could cause serious spectral 'contamination' (especially if high binding energy features such as O 2s and F 2s are present, see below); (3) the problems of sample damage during data acquisition are minimised (see Section 3.3). With modern instruments high resolution spectra can be acquired with good statistics in 1–2 hours and surface potential stability concerns are effectively eliminated. Such spectra from over 100 polymers are available in Beamson and Briggs (1992a).

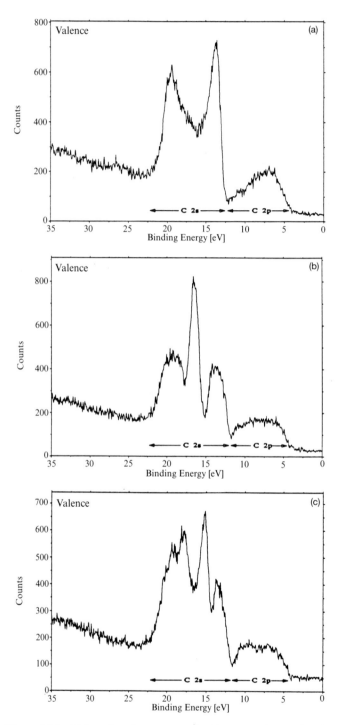

Figure 3.14   Valence band spectra, excited by monochromatic Al $K\alpha$ radiation, of: (a) high density poly(ethylene); (b) poly(propylene) and (c) poly(but-l-ene) (Beamson & Briggs, 1992a).

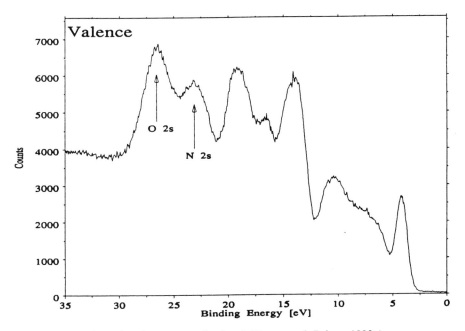

Figure 3.15   Valence band spectrum of nylon 6 (Beamson & Briggs, 1992a).

The spectra in Fig. 3.14 divide into two regions associated with orbitals derived from C2s (and H1s) and C2p (and H1s) atomic orbital mixing. In polymers containing heteroatoms such as O, N and F distinctive features representing the 2s level appear in the higher binding energy part of the spectrum, whilst the lower binding energy region becomes a rich structure of overlapping peaks resulting from the many additional orbitals formed from interaction of H1s, C2s/2p and, principally, O2p (etc) orbitals. This is illustrated in Figs. 3.15 and 3.16.

A particular study has been made of O2s binding energies and peak shapes for polymers containing carbon, hydrogen and oxygen (Briggs & Beamson, 1993). It is important to note that this region can be split by symmetrised combinations of O2s orbitals leading to molecular orbitals with different energies. Thus, for instance, in carboxylates, the O2s region has two equal components due to bonding (high binding energy) and antibonding orbitals separated by ~2.5 eV. These splittings can be qualitatively predicted by simple molecular orbital theory and local symmetry approximations. Similarly, an isolated CF_3 group will give rise to an F2s region split into a 2:1 doublet (analogous to the 3:1 splitting in CF_4 due to a and t combinations of F2s atomic orbitals (Gelius, 1974)). Both these F2s and O2s splittings are seen in Fig. 3.16.

From the above it is clear that valence band spectra have the potential to aid structural studies greatly. Currently, however, their usage is at the 'fingerprint' level even for pure polymers. To exploit this information level fully, theoretical

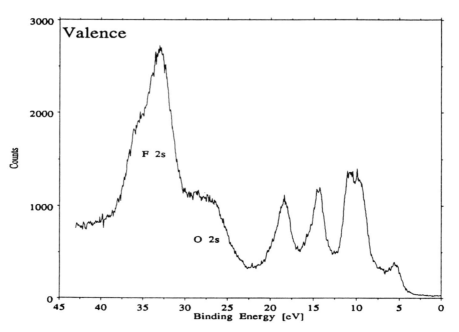

Figure 3.16 Valence band spectrum of poly(trifluoroethyl acrylate) (Beamson & Briggs, 1992a).

calculations are essential. Over the years several approaches to these have been used. Examples of band structure calculations and X alpha calculations can be found in Chtaib *et al.* (1991) and Xie & Sherwood (1993), respectively.

3.7   Auger series

The Auger series excited by $Mg\,K\alpha$ and $Al\,K\alpha$ of relevance to polymer studies are the KVV transitions of carbon, nitrogen, oxygen, fluorine and sodium. In the spectrum of gas phase sodium several sharp peaks associated with various KLL transitions are seen, of which the most significant are those for $KL_1L_{2,3}$ and $KL_{2,3}L_{2,3}$. As discussed in Section 2.2.1.3, when these first row elements form covalent bonds the L levels (particularly $L_{2,3}$) are involved in molecular orbitals and the term KVV series is more appropriate. In effect the valence band structure discussed above becomes convoluted into the Auger transitions leading to a broadened spectrum. For carbon KVV two more-or-less obvious regions are observed, related to the $KL_1L_{2,3}$ and $KL_{2,3}L_{2,3}$ peaks seen in the free atom spectrum. The latter is the more intense and at lower binding energy. For nitrogen, oxygen and flurorine a third component to higher binding energy of these two, related to KL_1L_1 becomes more obvious (Fig. 3.17). The approximate positions of the series maxima on the binding energy scale, under $Al\,K\alpha$ radiation, are

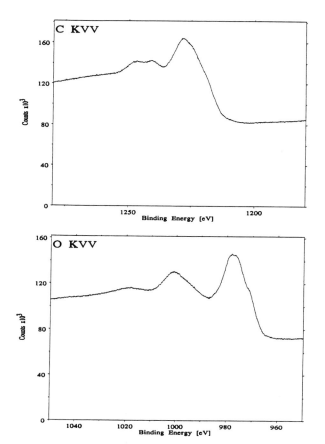

Figure 3.17 C *KVV* and O *KVV* Auger series from the Al *Kα* excited spectrum of PVMK (Beamson & Briggs, 1992a).

respectively: carbon (1225 eV); nitrogen (1110 eV); oxygen (975 eV): fluorine (830 eV); sodium (497 eV).

When Mg *Kα* radiation is used these energies remain the same whereas all the photoelectron peaks move by 233 eV to lower binding energy (on the kinetic energy scale the reverse is true). In this situation the Na *KLL* series overlaps the C 1s region which can cause confusion. The *KVV* spectra themselves are not used analytically in polymer XPS.

3.8   Functional group labelling (derivatisation)

From the discussion in Sections 3.4.4–3.4.6 it is obvious that many functional groups give core level binding energies (or chemical shifts) which are non-specific. The resulting ambiguity when attempting interpretation of polymer spectra is even more pronounced for multifunctional surfaces because of the general broadening of peaks due to (smaller) secondary shifts and subtle variations in local structure around any particular functionality.

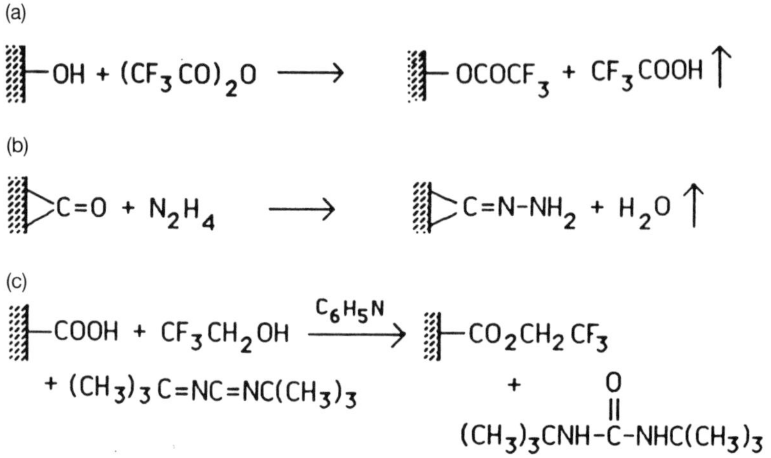

(a)

$$\text{—OH} + (CF_3CO)_2O \longrightarrow \text{—OCOCF}_3 + CF_3COOH \uparrow$$

(b)

$$\text{>}C\text{=}O + N_2H_4 \longrightarrow \text{>}C\text{=}N\text{-}NH_2 + H_2O \uparrow$$

(c)

$$\text{—COOH} + CF_3CH_2OH \xrightarrow{C_6H_5N} \text{—CO}_2CH_2 CF_3$$

$$+ (CH_3)_3 C\text{=}NC\text{=}NC(CH_3)_3 \qquad + \qquad \underset{\parallel}{\overset{O}{(CH_3)_3CNH\text{-}C\text{-}NHC(CH_3)_3}}$$

Figure 3.18 Vapour phase derivatisation schemes for labelling surface functional groups: (a) hydroxyl by reaction with trifluoroacetic anhydride; (b) carbonyl by reaction with hydrazine and (c) carboxylic acid by reaction with trifluoroethanol in the presence of pyridine and di-t-butyl carbodimide (Chilkoti, Ratner & Briggs, 1991a).

This frustration has led to a significant effort to develop derivatisation schemes whereby a specific derivatising agent reacts with a single functional group and labels it with a distinctive element. This procedure may have the additional advantage of increasing the detection sensitivity if the elemental 'tag' has a higher cross-section than C1s, O1s, or N1s. Many such reactions have been studied and a fairly complete listing, with original references, can be found in Briggs (1990b).

Besides being specific, the derivatising reaction should proceed rapidly under mild conditions and go to completion within the volume sampled by XPS. As research in this area has proceeded it has, unfortunately, become clear that these ideal conditions are almost impossible to achieve. Early work involved solvent-based reactions and it was frequently found that reactions which are rapid at room temperature in solution are either inhibited in the polymer surface layers or may even follow a different course (e.g. E2 elimination instead of S_N2 nucleophilic substitution), probably because of steric effects. Solvents which aid the reaction by swelling the polymer are likely to promote surface reorganisation, e.g. functional groups specifically at the surface migrate into the bulk. Gas or vapour phase reagents minimise these latter difficulties but they are generally much more difficult to identify. Perhaps the greatest problem is that of specificity. Few of the schemes in the literature have been subjected to a detailed study of selectivity; thus the existence of a scheme in the list mentioned above does not signify that the necessary selectivity has actually been proven. When this is thoroughly tested the results are frequently disappointing. The selectivity of vapour phase reagents for derivatising oxygen functionalities, including those illustrated in Fig. 3.18, has been studied by Chilkoti *et al.* (1991a,b) and these papers provide a most detailed insight into the practice and problems of derivatisation.

3.9 Non-destructive depth-profiling

This technique is based on the variation of the vertical sampling depth with photoelectron take-off angle, discussed in Section 2.2.2, and is often referred to as angle-resolved XPS (ARXPS).

It is a particularly important technique for polymer surface studies because destructive depth-profiling by argon ion etching (sputter depth-profiling) is inappropriate. Under these conditions polymers rapidly degrade to materials not unlike (doped) amorphous carbon. Unfortunately, however, the depth regime which can be studied is limited to <10nm as shown in Table 2.3. Nevertheless, many important aspects of polymer surface behaviour are governed by major compositional or structural variations within this layer thickness as discussed in Chapter 1. At the semiquantitative or merely qualitative level significant insights may be gained from studying the surface at two extreme take-off angles, e.g. 10° and 90°. The lowest angle that can be used may be constrained by instrumental geometry considerations and/or by signal:noise considerations (in many instruments sensitivity drops significantly at low take-off angle). Two examples of ARXPS are given in Figs. 3.19 and 3.20.

In principle, it should be possible to obtain spectra for a series of take-off angles (typically four or five data sets) and then reconstruct the concentration depth-profile since, as discussed in Section 2.2.2:

$$I_A = K \int_0^\infty C_A(z) \exp(-z/\lambda_A \sin\theta)\, dz \tag{3.6}$$

where I_A is the intensity of a peak representing element A of interest, $C_A(z)$ is the concentration of A at depth z, λ_A is the attenuation length for an electron of the appropriate kinetic energy, θ is the take-off angle (to the sample surface) and K is a normalisation parameter.

However, equation (3.6) is representative of a mathematically well-known class of 'ill-posed' problems, the solution of which requires a matrix inversion for which there is no unique solution. Moreover, the solution (i.e. the concentration depth-profile) can be extremely sensitive to small variations in input parameters. Despite these problems, several algorithms have been proposed and used for calculating depth-profiles and layer thicknesses. A detailed comparison has been made by Tielsch and Fulghum (1994). In all ARXPS work it is important to use as small an analyser acceptance angle as possible in order to define the take-off angle accurately.

Two final cautionary notes need to be sounded. Firstly, equation (3.6) is based

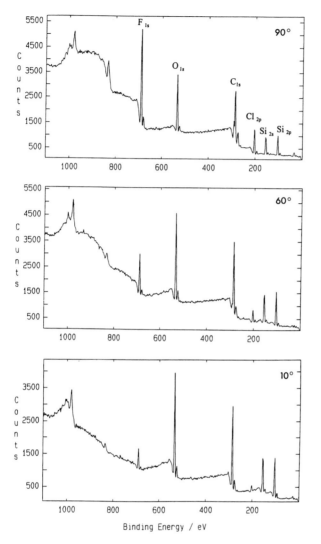

Figure 3.19 Example of surface contamination investigation by ARXPS. Spectra from the surface of a plaque of poly(ethylene–chlorotrifluoroethylene) contaminated with poly(dimethylsiloxane) (dimethylsilicone) taken at 90°, 60° and 10° from the surface. Note the rapid attentuation of the substrate signals (F 1s, Cl 2p) with increasing surface sensitivity (decreasing take-off angle).

on the incorrect assumption that elastic scattering effects can be neglected (see Section 2.2.2). The complications of elastic scattering effects are only now being evaluated. Secondly, the surface is taken to be atomically flat. Roughness effects on ARXPS were discussed at an early stage by Baird & Fadley (1977). As atomic force microscopy (AFM) studies of polymer surfaces become more widespread, it is clear that many apparently (or assumed) flat polymer surfaces which give rise to significant spectral changes with ARXPS are quite rough on the scale of electron attenutation lengths. Therefore, the analysis of ARXPS data requires some caution and it is unlikely that quantitative interpretation based on simple models will be valid.

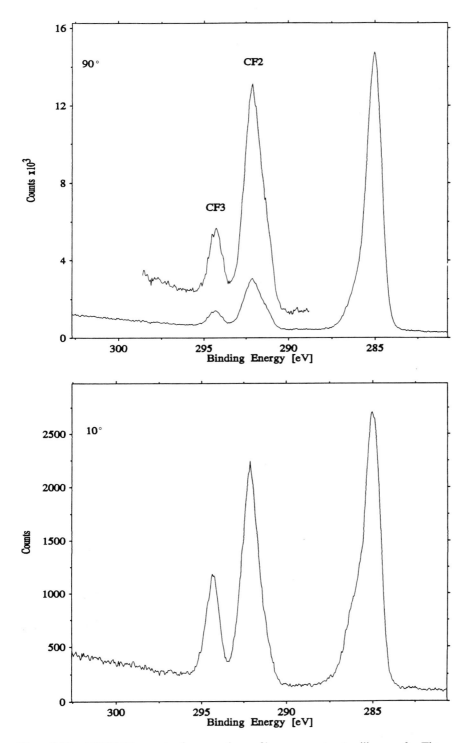

Figure 3.20 ARXPS C 1s spectra from a polymer film spun-cast onto silicon wafer. The polymer has a hydrocarbon backbone onto which are grafted pendant linear chains ending in C_6F_{13} units. The strong enhancement of the CF_2/CF_3 peaks relative to CH_x (at 285 eV) at low take-off angle indicates a high degree of order and alignment of the C_6F_{13} chains (perpendicular to the plane of the film). At 90° the sampling depth is of the same order as the length of the C_6F_{13} unit and the CF_3/CF_2 peak intensity ratio (0.22) is close to the stoichiometric value (0.20). At 10° this ratio increases to 0.36 as the sampling depth has been greatly reduced, so weighting the intensity of the peak due to the outermost CF_3 group.

Chapter 4

Static SIMS (SSIMS)

4.1   Instrumentation

SSIMS involves the bombardment of a sample with a low density flux of positive ions (or neutral atoms) and the mass analysis of the positive and negative *secondary* ions which are emitted from the sample surface. A secondary ion mass spectrometer (or SIMS instrument) therefore consists of a vacuum vessel with its associated pumping system and sample introduction/manipulation systems; a primary ion (or atom) source; a mass spectrometer and its associated secondary ion collection optics; a secondary ion detection system and a dedicated 'datasystem' based on a PC or workstation for control of the spectrometer and processing of the acquired data. In addition, when studying insulating samples such as polymers it is necessary to overcome charging problems by use of an auxilliary source of electrons. Thus an electron source is an essential component of the SSIMS instrument. These components are now considered in turn.

4.1.1   Vacuum system and sample handling

The construction of the stainless steel vacuum system, the options for pumping systems and the sample transfer/manipulation mechanisms are essentially identical to those described for XPS instruments (Section 2.1), except that automation of sample positioning/data acquisition from multiple samples is not yet routine in SSIMS (see Section 4.2.7). Operation in the UHV regime is even more important in SSIMS than in XPS. SSIMS is inherently more surface sensitive so that surface contamination by adsorption causes more problems; not only is the available surface for analysis reduced but also the spectrum of the contaminating material has a much greater impact on the acquired spectrum in SSIMS than would be the case for the equivalent XPS experiment.

4.1.2   Primary ion sources

A variety of ion gun types is used in SSIMS (Jede, Ganshow & Kaiser, 1992) and these give rise to the range of primary ion/energy combinations which are encountered. This variety is due to the different requirements for guns which allow work at high or low spatial resolution in addition to the constraints imposed by the type of mass spectrometer employed (either quadrupole, requiring a continuous beam, or time-of-flight (ToF), requiring a pulsed beam). The most important ion source parameters are brightness, extractable current and energy spread. These parameters, given the design philosophy and quality of the ion-optical column for transporting, focusing and possibly pulsing the ion flux, will determine the final current/spot size characteristics of the beam at the sample surface. The three types of ion source used for SSIMS are the electron impact, surface ionisation and liquid-metal field emission sources.

In the electron impact ion source, electrons from a heated filament (cathode) are accelerated towards an anode by a voltage difference of the order of 100–200 V and thus gain sufficient energy to ionise supply gas atoms on impact. The design of the source region and the associated electrodes is such as to produce multiple passes of the electrons through the ioniser thus increasing the extractable ion current. Because most of the ions are formed within an equipotential the energy spread is 5–10 eV. These sources usually operate with noble gases (reactive gases, e.g. O_2, N_2, lead to rapid filament failure) at a pressure of 10^{-7}–5×10^{-4} mbar. Typically the energy is variable from 0.1–5 keV allowing spot sizes from $\sim 50 \, \mu m$ to several millimetres. With a beam brightness of the order of $10^{-2} \, A \, cm^{-2} sr^{-1}$ the maximum current density at the sample is of the order of $1 \, mA \, cm^{-2}$. Ion guns based on this type of source are small, relatively cheap and ideal for quadrupole SSIMS. The necessary stable, low current density at the sample ($<1 \, nA \, cm^{-2}$) is easily achieved by using a low gas pressure and a defocused beam and/or raster scanning over an area several mm square. Typically beams of 2–4 keV Ar^+ or Xe^+ are employed. More sophisticated designs may incorporate a Wien (mass) filter to ensure an isotopically pure ion beam but this is unnecessary for quadrupole SSIMS of polymers.

An electron impact source has also been used in a pulsed ion gun for ToF SIMS. This gun delivers 10 keV Ar^+ ions in pulses of 0.8 ns half-width with a spot diameter of 4–10 μm and 500–1000 primary ions per pulse. The deflection arrangement for pulsing the beam simultaneously provides the necessary mass filtering in this design (Schweiters et al., 1991).

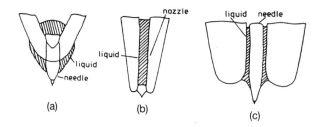

Figure 4.1 Schematic representation of the different types of liquid-metal ion sources:
(a) needle/filament type;
(b) nozzle type;
(c) needle/nozzle type
(Jede *et al.*, 1990).

Surface ionisation sources produce Cs^+ beams used for ToF SIMS. These sources can have very high brightness (up to $500\,A\,cm^{-2}\,sr^{-1}$) combined with a very low energy spread. In this type of source caesium is fed from a heated reservoir either into a heated porous tungsten plug or, via the vapour phase, onto a heated filament. In either case evaporation of caesium from the tungsten surface occurs both as atoms and ions. If the rate of desorption exceeds the rate of arrival of atoms at the surface the ratio of ion/atom desorption increases and, in practice, almost 100% ionisation can be achieved. These ions are then accelerated away from the emitting surface. Since no collisions are involved the ion beam is very pure and since evaporation is by thermal means the energy spread is $\sim 2kT$ (0.2 eV). The high purity of the beam means that mass-filtering is not required and the high brightness ensures that more ions per pulse can be achieved than with an electron impact source. Typically these sources are used to deliver relatively large spot sizes ($100\,\mu m$ or greater either for spectroscopy or for stigmatic (i.e. non-microprobe) imaging) although reasonable current densities are still possible when focusing to $<5\,\mu m$ for medium resolution microprobe imaging. A typical extraction voltage is 8 kV.

Liquid-metal ion sources have been developed in particular to advance ion microprobe performance. In a typical source (Fig. 4.1) the liquid metal (usually gallium) from a (heated) reservoir is drawn over the tip, radius $\sim 5\,\mu m$, of a needle. In front of the tip is an extraction electrode biased negatively so as to produce a field strength at the tip of $\sim 10^8\,V\,cm^{-1}$. Opposing electrostatic and surface tension forces acting on the liquid film produces a conical shape (Taylor cone) with a very high radius of curvature cusp ($\sim 2\,\mu m$) protruding from the tip. From this cusp field ion emission occurs. The source brightness is exceptionally high, $\sim 10^6\,A\,cm^{-2}\,sr^{-1}$, but the energy spread is relatively large and depends on the extracted current (5–35 eV for currents of 1–25 μA). Optimum performance for high spatial resolution is therefore achieved at the lowest possible ion current (1 μA) and at high extraction voltage (25–30 kV). Under these conditions sample current densities of $1\,A\,cm^{-2}$ in spot diameters of $>30\,nm$ are possible.

Pulsed liquid-metal ion source guns are widely used for high resolution ToF SIMS imaging since they are the only type which can readily realise spot sizes well below 1 μm. Pulsing the ion beam inevitably leads to spot size degradation relative to the continuous beam and the final resolution is typically 0.1–0.2 μm if the pulse length is very short (1 ns), as required for obtaining high mass resolution (Section 4.1.5). As the pulse length increases so does the potential spatial resolution. This has led to schemes for variable beam bunching so as to provide some choice of mass resolution/spatial resolution/sensitivity combination.

Gallium has two isotopes: ^{69}Ga (60.1%) and ^{71}Ga (39.9%). Early liquid-metal ion sources used natural gallium and therefore required a mass separation stage in the gun column (itself a 'ToF'). Current designs use isotopically separated ^{69}Ga which avoids this problem and produces fractionally greater current at the sample. It seems likely that indium, which like gallium has a conveniently low melting point, may soon be used in LMIS for SSIMS. Its two isotopes are ^{113}In (4.3%) and ^{115}In (95.7%). In this case the use of the natural metal with a mass filtering gun column is much more efficient than the gallium equivalent.

4.1.3 Atom sources

The charging problem (see Section 4.2.7) inherent in SSIMS studies of insulators led to the development of fast atom sources. This removes the deposition of positive charge which is thought to be more important to the charging mechanism than secondary electron emission. An ion beam from a noble gas electron impact source is passed through a cell containing the same gas at high pressure ($\sim 10^{-3}$ mbar). This leads to charge (electron) exchange between the fast ion and the relatively stationary gas atom. There is virtually no momentum transfer so that the fast atom beam which emerges from the cell has the same energy and spatial characteristics as the original ion beam. Commercial sources merely produce collimated beams of millimetre dimensions. Relative to noble gas ion sources there are both advantages and disadvantages associated with using atom sources. Whilst charging in positive ion mode is sufficiently reduced to avoid the need for electron flood neutralisation, the latter is still required in negative ion mode. Measurement of atom beam density is more difficult and the beam cannot be steered. However, there is growing evidence that sample damage rates are significantly reduced with atom bombardment compared with ion bombardment at equivalent fluxes. Sources which combine the facilities for both ion and atom beam production are available (Brown, van den Berg & Vickerman, 1985).

4.1.4 Quadrupole mass spectrometers

The quadrupole mass filter (Fig. 4.2) consists of four electrically conducting parallel rods. Opposite rods are connected to the same potential, which has both a dc and an rf component, and for each pair of rods the sign of this potential is opposite, i.e.

$$P(t) = \pm [U + V\cos(2\pi f t)]$$

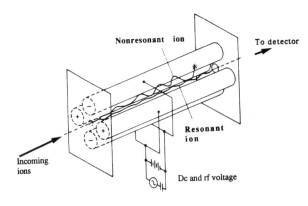

Nonresonant ion

To detector

Resonant ion

Incoming ions

Dc and rf voltage

Figure 4.2 Electrical connections and secondary ion trajectories between the rods of a quadrupole mass analyser.

where U is the dc voltage and V the peak amplitude of the rf voltage at frequency f.

Ions injected into this assembly parallel to the rods undergo transverse motion leading to oscillatory trajectories mainly leading to collision with the rods. For certain values of U, V and f any ion of given m/z can have a stable trajectory through the assembly. There are several ways in which these parameters can be systematically varied to scan a mass range, leading to different interrelationships between mass resolution and transmission. For SSIMS, quadrupoles are usually set up to give Δm (width of peak) constant at ~ 1 amu for any ion; consequently transmission, T, decreases with secondary ion mass as roughly $T \alpha m^{-1}$. Overall transmission is increased by using larger diameter rods and the higher the frequency f (usually fixed) the shorter the rods can be.

Two other factors can significantly affect quadrupole transmission properties. Large fringeing fields exist at the ends of the rods leading to trapping and reflection of low velocity ions. An additional set of short rods at the main quadrupole entrance (leading to a so-called segmented rod configuration) overcomes this problem and greatly improves transmission. The quadrupole cannot filter ions of high kinetic energy and so it is preceded by some form of energy filter, typically accepting ions of *ca* 10 ± 5 eV. However, since secondary ion kinetic energy distributions can vary significantly (for instance cluster ions have a much narrower distribution, peaking at lower energy, than atomic ions) the settings of the energy filter (i.e. energy and band pass) play a dominant role in determining the relative intensities of peaks, particularly between those due to atomic and molecular fragment ions. Finally, some form of extraction lens is used to collect the secondary ions and inject them into the energy filter/quadrupole. For SSIMS it is usual to employ low extraction fields and high angular acceptance from large fields of view, although higher extraction energies, leading to smaller fields of view, have been employed to improve transmission. This necessitates a further decleration step prior to injection into the quadru-

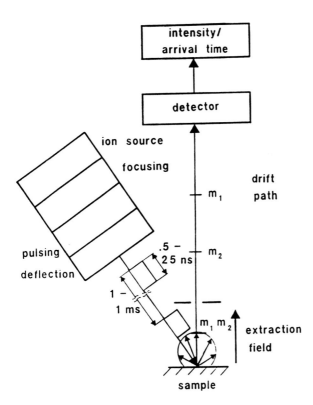

Figure 4.3 Schematic of a ToF SIMS instrument (Jede *et al.*, 1992).

pole assembly as noted above. Depending on the form of these collection optics, a 'large' quadrupole will have a transmission (ratio of ions detected to those emitted from the sample) of 0.1–1%. Such a quadrupole will have a mass range of between 1 and ≈800 amu. Typically, a polymer spectrum would be acquired with a 2–4 keV Ar^+ or Xe^+ primary ion beam of spot size ≈100 μm rastered over an area several millimetres square with a current density of ~1 nA cm^{-2}. Adequate intensity requires a total dose per spectrum of $(1–5)\times10^{12}$ ions cm^{-2} (depending on secondary ion yields). In this context, the aspect of quadrupole mass spectrometry to emphasise is the scanned mass range. Whilst the sample is being continuously dosed with primary ions only a fraction of the total spectrum is being recorded at any point in time. As discussed in Section 4.2.6 there is a crucial requirement to limit the total ion dose during SSIMS analysis and, therefore, this is an inherently inefficient approach.

4.1.5   ToF mass spectrometers

A schematic of a ToF SIMS instrument is given in Fig. 4.3. A primary ion pulse produces a 'packet' of secondary ions which are accelerated to 3–5 keV over a very short distance, thereby giving them all virtually the same kinetic energy (E_K) before entering a field-free drift tube. Since $E_K = \frac{1}{2}mV^2$ (for singly charged ions),

ions of different mass will have different velocities and mass separation will occur. For a linear drift tube of flight path L

$$t-t_0=L(m/2E_K)^{1/2} \tag{4.1}$$

where t_0 is the start time and t the arrival time of the singly charged secondary ion with mass m. A typical value for L is 2 m giving an acceptable flight time of $\approx 200\,\mu s$ for the highest mass ions which can practically be detected ($m/z\sim 10000$). This time, plus a similar time for processing the data, must elapse before the next primary ion pulse, so the repetition rate is in the kilohertz range.

Mass resolution is given by:

$$\frac{m}{\Delta m}=\left\{\frac{(2\Delta t_1)^2}{t}+\frac{(2\Delta t_A)^2}{t}+\frac{(2\Delta t_D)^2}{t}\right\}^{-3/2} \tag{4.2}$$

where Δm is the peak width, Δt_1 is the time spread due to the finite width of the primary ion pulse, Δt_A that due to time focusing aberrations in the analyser and Δt_D that due to rise- and deadtime in the detector and its electronics, t being the total flight time. Minimum pulse lengths are $\sim 0.5\,ns$. The major consideration for the ToF analyser is the initial energy spread of the secondary ions which will give a range of t for any value of m. Three designs have been used to overcome this inherent problem with a linear ToF analyser. In the 'Poschenrieder' analyser a toroidal electric field produced by electrostatic sectors is interposed between two linear drift tubes. In this field the secondary ions follow curved trajectories which are longer for the more energetic ions in such a way that the initial energy variation is compensated (Oetjen & Poschenrieder, 1975). In the 'reflectron' analyser (Mamyrin et al., 1973) an ion mirror performs a similar function (Fig. 4.4) and is capable of maintaining higher mass resolution at high transmission than the Poschenrieder design. This is now a popular choice. Finally, the so-called TRIFT analyser, originally introduced for stigmatic imaging (described in Section 4.1.7) is also capable of high mass resolution by energy compensation. Modern ToF instruments can achieve $m/\Delta m$ of >7000 at m/z 28 and >10000 at m/z 500 and still have mass-independent transmission of $>20\%$, approaching 100% for the large organic fragments typical of polymer SIMS.

In contrast to the quadrupole situation, the ToF approach is ideally suited to SSIMS. The whole spectrum is acquired in parallel so the primary ion dose is the minimum possible. Transmission is near-maximum, there is no discrimination against high mass ions (at least to several thousand in m/z) and, for most practical purposes, no upper mass limit. The area analysed in a ToF SIMS instrument is typically $\leqslant 200\,\mu m$ square (in order to obtain high transmission), which is

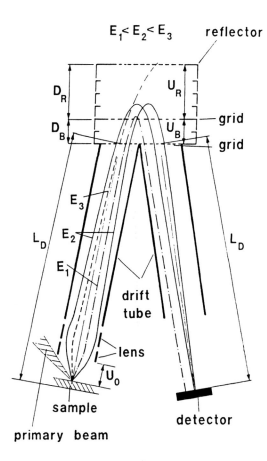

$E_1 < E_2 < E_3$ reflector

D_R

U_R

D_B U_B — grid

— grid

E_3

L_D E_2 L_D

E_1

drift
tube

lens

U_0

sample

detector

primary beam

Figure 4.4 Schematic of a two-stage reflectron ToF mass analyser showing the trajectories of three secondary ions with the same composition but different initial energies (Jede *et al.*, 1992).

significantly less than in a quadrupole instrument. However, this is more than compensated for by the large gain in sensitivity (of the order of 10^4) and spectra can be obtained for doses of $<10^{11}$ ions cm^{-2}.

4.1.6 Detection systems

For quadrupole instruments used under SSIMS conditions channel electron multipliers (see Section 2.1.6) operating in the pulse counting mode are employed. Secondary ions are accelerated into the channel electron multiplier with an energy of kiloelectron volts, the value depending on ion polarity.

ToF instruments employ chevron stacks of channel plates (see Section 2.1.6). In early instruments the output was grounded or capacitively decoupled from high potential. Modern instruments operating at high mass resolution employ channel plates followed by a resistive anode encoder. Large slow moving ions with an *m/z* of the order of several thousands tend to give low electron conversion yields on impacting the channel plate. Variable post acceleration (to 10 keV or more) into the detector has been introduced to reduce this problem. However, the detector set-up, particularly pulse height discriminator threshold, can have a major impact on high mass peak intensity.

4.1.7   Imaging SSIMS

For quadrupole instruments imaging can only be carried out using the micro-probe approach, i.e. the primary ion beam is focused to produce a small spot at the sample and the beam is raster-scanned (digitally) over the surface. The mass filter is tuned to detect a signal (secondary ions of a chosen $\pm m/z$), representing a species of interest, whose intensity at each pixel in the scanned array is mea-sured. The counting (dwell) time of the beam at each pixel will depend on the peak intensity. An early example of polymer imaging is given in Fig. 1.3. Imaging multiple species requires each image to be acquired serially with an accumulat-ing ion dose throughout.

Early experience showed that the spatial resolution available from liquid-metal ion guns (i.e. $<1\,\mu m$) could not be effectively utilised, in general, for polymer surfaces because of the sensitivity limitations of the quadrupole mass spectrometer and/or the charging problems associated with use of a fine-focus beam (see Section 4.1.8).

For ToF instruments both microprobe and microscope approaches are avail-able. In the former case, the advantage of the ToF mass spectrometer is that the whole spectrum is acquired at each pixel. In early instruments a certain number of masses (or groups of masses) could be chosen prior to commencing the imaging experiment; one scan over the selected area yielded multiple images. The restriction was basically one of computer memory (Briggs et al., 1990). Today, there are no such restrictions so selected ion imaging can be 'retrospective' fol-lowing the acquisition of just two datasets (the complete positive and negative ion spectra at each pixel).

Retrospective image analysis (Reichlmaier, Bryan & Briggs, 1995) is illus-trated in Fig. 4.5. The complete dataset is first condensed into two files: the first is the spectrum obtained by summing the spectra in all pixels (average spectrum) and the second is the image obtained by summing the spectral intensity in each pixel (total ion image). Peaks in the summed spectrum can be selected (singly or collectively) and an image created based on their intensity in each pixel. Alternatively, a group of pixels can be selected (bounded) and the spectra in each summed. This is restrospective microanalysis.

A ToF microscope employing a stigmatic imaging analyser is shown schematically in Fig. 4.6. (Schueler, Sander & Reed, 1990). The sample is broadly illuminated with a Cs^+ beam and an immersion lens forms a secondary ion image which is magnified by projection lenses. Diaphragms define the angular acceptance (spatial resolution) and field of view after which a series of three 90° electrostatic sectors with intermediate drift spaces forms the image onto a position sensitive detector. Focusing is independent of secondary ion initial energy with respect to flight time (mass) and lateral position. The whole image is, therefore, acquired in parellel. The same instrument can operate as a microprobe when a scanned, focused beam is used. This gives higher spatial resolution and a higher depth of field but sacrifices the advantage of image acquisition time.

4.1.8   Electron sources for charge compensation

The combination of positive charge injection (primary ions) and secondary electron emission leads insulating samples to charge up positively during SSIMS analysis. This is discussed in more detail in Section 4.2.8. This charging is compensated or, more accurately, the sample surface potential is appropriately stabilised, by means of simultaneous electron bombardment.

In the case of quadrupole SIMS simplified versions of electron guns designed for AES have been used. These allow defocused broad beams (many millimetres spot size) of up to several kiloelectron volts energy to be produced and steered into position. Typically, the energy used is 50–1000 eV with a current density in the static beam of the same order as the primary ion current density, or slightly larger.

In the case of ToF SIMS the electron sources are designed to give high fluxes of low energy (10–70 eV) electrons. However, they also have to be pulsed so that the electrons are sprayed onto the sample during the period of secondary ion flight time analysis, when the extraction field is switched off.

4.1.9   Datasystems

Datasystems for quadrupole SSIMS instruments are relatively uncomplicated and based on computers in the PC category. They minimally control the mass range scanning, the dwell time at a particular m/z interval (and therefore the count rate) and the acquisition of the spectrum. In depth-profiling the mass spectrometer can be rapidly switched between masses of a single polarity in order to record intensity for a series of peaks as a function of time/ion dose. The degree to which all instrument voltages can be digitally controlled will depend on how modern the system is. Processing capabilities include spectral smoothing, normalisation, addition and subtraction.

Modern ToF SIMS instruments require powerful datasystems. These are based on a workstation computer and may have subsidiary PCs to control specific functions. Virtually all aspects of the instrument operation are under digital voltage control. The datasystem is, of course, interfaced with the fast electronics which control the complex timing sequence of primary ion and

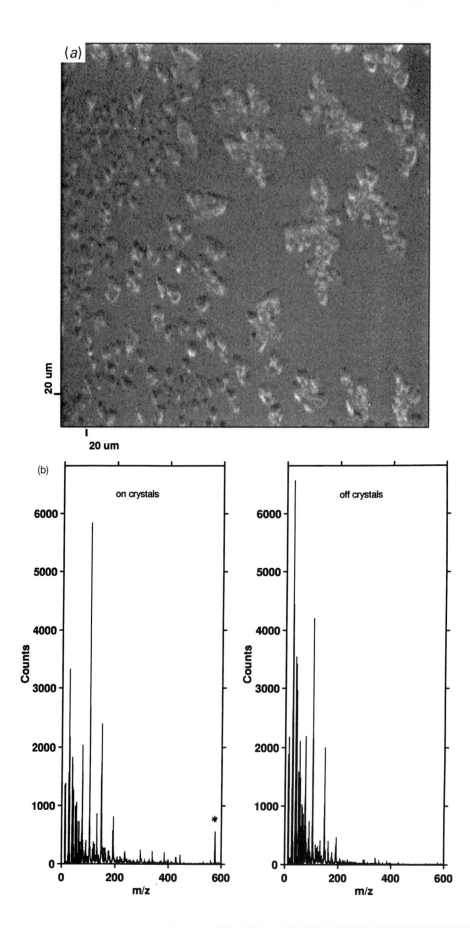

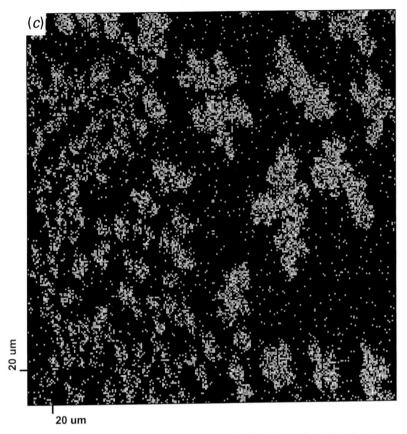

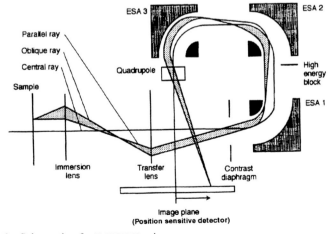

Figure 4.5 An example of retrospective image analysis. The total positive ion image (a) of cyclic trimer crystals on the surface of PET film is dominated by topographic contrast. The spectra (b) correspond to the defined regions which are mainly 'on' the crystals (sum of four such areas) and 'off' the crystals. Since the fragmentation patterns of the polymer and cyclic oligomer are almost identical, it is the detection of the trimer molecular ion, $[M+H]^+$ at m/z 577 (*), which distinguishes the latter. Constructing a retrospective image based only on the intensity of the m/z 577 peak (c), reproduces the information in the total ion image but based on chemical contrast. The maximum number of counts per pixel in images (a) and (c) is 339 and 6 respectively (Reichlmaier *et al.*, 1995).

Figure 4.6 Schematic of a ToF SIMS microscope.

electron flood pulses and extraction field switching and which determine the secondary ion flight times. The computer generates the mass spectrum from the spectrum of ion arrival times and performs internal mass calibration. Imaging involves, additionally, the sorting of spectra with pixel position which is related to either ion beam co-ordinates (microprobe) or position sensitive detector co-ordinates (microscope). Storage of this information requires very large memory capacity. Image processing software is now very advanced, especially in terms of enhancing contrast which is important for the low counts per pixel situations usually encountered in polymer imaging. 'Retrospective' analysis was discussed in Section 4.1.7.

4.2   Physical basis

In contrast to XPS, discussed in Chapter 2, the physical basis of SSIMS is rather poorly understood. This statement is true in general and especially so for SSIMS of polymers. Although there is no question that the mass spectrum obtained strongly reflects the molecular structure of the surface, the way in which the detected secondary ions (and their relative intensities) arise is still a matter of considerable conjecture. The two process involved in the production of secondary ions are *sputtering* and *ionisation*. Whether these occur simultaneously or consecutively is one of the debated issues. However the ions leaving the surface region are formed initially, they may *fragment* during their flight to the detector.

4.2.1   Sputtering

Sputtering is the process by which secondary particles are emitted from the surface as a result of high energy primary particle impact. A number of mechanisms may be involved; these have different time-scales and their relative importance depends both on the primary particle energy and flux (particles $cm^{-2}s^{-1}$). At low energy ($<1\,keV$) and low flux, the so-called knock-on regime, atoms are ejected from the surface layer by direct impact if they have sufficient energy to overcome the surface binding forces. Only the single target atom is displaced on impact (i.e. only primary recoil events are involved). The process is called *prompt collisional sputtering* and it occurs within 10^{-15}–10^{-14}s of impact. At higher energies and fluxes the recoil atoms have sufficient energy to generate secondary and higher generation recoils (the linear cascade regime). However, knock-on collisions still dominate and collisions between moving atoms are infrequent. Sputtering can therefore occur some distance from the point of impact. The process is called *slow collisional sputtering* and it occurs within 10^{-14}–10^{-12}s of impact. Finally, at even higher flux, most of the target atoms within a certain volume are moving as a result of the very high density of recoil atoms and

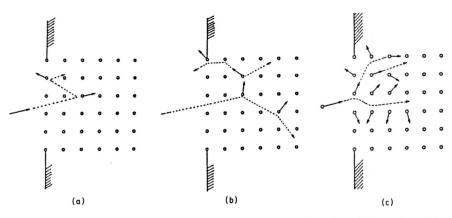

Figure 4.7 Schematic of sputtering by elastic collisions: (a) single knock-on regime; (b) linear cascade regime; (c) spike regime (Sigmund, 1981).

sputtering may be likened to transient vaporisation (the thermal spike regime). Although referred to as *thermal sputtering* there may be a significant contribution from electronic sputtering due to the high degree of electronic excitation in this process. The timescale is $10^{-13}-10^{-10}$ s. These three regimes are depicted in Fig. 4.7.

The collision cascade regime is of most relevance under normal SIMS conditions. The collision cascade theory of sputtering, developed mainly by Sigmund (Sigmund, 1981) has been quantitatively successful in predicting atomic sputter yields (number of sputtered atoms per incident primary ion) and secondary particle energy distributions. It is based on linear transport theory and billard-ball-like collisions. This has been important for understanding the *dynamic SIMS* experiment in which relatively high sputtering rates are used either to detect low concentrations of elements in bulk or, more specifically, to determine their concentration as a function of depth below the surface (Wittmaack, 1992). The process has also been modelled by Monte Carlo simulations and an early example is shown in Fig. 4.8. This nicely illustrates the shallow sampling depth of SIMS (~1 nm) despite the much greater depth to which the cascade extends, and the lateral displacement of sputtered atoms (<5nm) from the impact point. However, two points need to be emphasised. First, the systems to which these theories/models have been applied are relatively simple (compared to polymers), such as metals and silicon with well-defined lattices. Second, the primary concern has been to understand the emission of atomic species. Even for these systems realistic molecular dynamics simulations have only just begun to give any insight into the emission of *cluster* ions (an assembly of atoms bearing an overall charge).

For primary particle energies in the kiloelectron volt range used in SSIMS, most of the energy is transferred to the sample by nuclear stopping collisions. This contrasts with the situation for much higher energy primaries (megaelectron volt range) where most of the energy transfer is through electronic

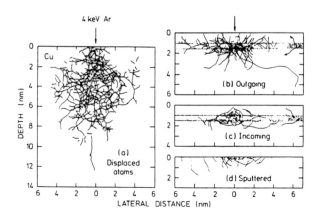

Figure. 4.8 Computer simulation of the relocation of copper atoms due to the impact of 4 keV argon ion: (a) trajectories within the whole collision cascade for 10 incident particles; (b), (c) transport of target atoms out of and into the marked layer (the uppermost 1 nm) for 20 incident particles; (d) trajectories of sputtered atoms for 50 incident particles (Ishitani & Shimizu, 1974).

interactions. Sputtering in this case is referred to as electronic sputtering. Such conditions are used in plasma desorption mass spectrometry (Vickerman, 1989a) which is capable of efficiently detecting very high mass (m/z>40 000) molecular ions. There is evidence to suggest that electronic effects may play a role in sputtering of insulators even under SSIMS conditions. This comes from comparisons of atom and ion bombardment under the same energy and flux conditions. The suggestion is that resonant – or Auger – neutralisation of the incoming primary ion can lead to electronically excited states in the insulator surface. If these are anti-bonding states then fragmentation and desorption may result (Leggett & Vickerman, 1992).

4.2.2   Ionisation

In general, the vast majority of the sputtered particle flux is neutral, with secondary ions comprising of the order of 1% (exceptionally this may rise to 10%). Ionisation probabilities for sputtered atoms can vary over four orders of magnitude. Related to this is the *matrix effect* in dynamic SIMS, whereby the ion yield for a particular atomic species depends on the chemical environment from which it is sputtered. For example, the secondary ion yield of Si^+ is about 70 times higher from silicon dioxide than from elemental silicon and that for W^+ about 400 times higher from tungsten trioxide than tungsten, under the same bombardment conditions. This same effect causes the major quantification difficulties in dynamic SIMS. A number of general models have been developed to explain the features of ion yields from metals and simple compounds such as oxides and ionic solids (Vickerman, 1989b). These have limited relevance for SSIMS of organic materials where the role of specific molecular structure in the process of cluster ion emission is clearly of crucial importance. However, all models of cluster ion formation share the same difficulty: at what point in the overall emission process does ionisation occur?

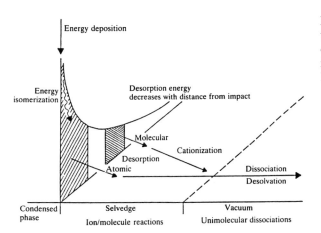

Figure 4.9 Summary of the processes thought to occur during desorption ionisation (Cooks & Busch, 1983).

4.2.3 Desorption ionisation

The desorption ionisation model (Cooks & Busch, 1983) specifically addresses cluster ion emission from organic materials. Here the desorption and ionisation steps are completely decoupled. Energy deposited by the primary particle eventually ends up being absorbed by molecules in small quanta to produce thermal/vibrational motion. 'Preformed' cations or anions (i.e. ionised species existing prior to impact) can be desorbed directly and pass through the selvedge region (the uppermost surface layers affected) and into vacuum without neutralisation. In the selvedge region desorbed neutral clusters may be ionised by attachment of small ions (H^+, metal ions, halide ions), by electron impact or via ion–molecule reactions. Finally, ions in the vacuum may fragment through unimolecular dissociation processes which are governed by the parent ion's internal energy and its distribution between the available electronic/vibrational states.

The desorption ionisation model is shown schematically in Fig. 4.9. A convenient picture of energy transfer involves two concentric circles, radii r_1 and r_2, centred on the point of impact. At distances, r, from this point where $r<r_1$, the energy transferred to molecules in the collision cascade is so high that many bonds are broken and only atoms or small fragments are emitted. At greater distances, where $r_1<r<r_2$, the average energy transferred is sufficient to desorb intact molecules or major fragments, a large fraction of which do not have enough internal energy to fragment in the vacuum. At even greater distances, where $r>r_2$, insufficient energy is transferred to cause desorption.

This model, though aimed at a understanding of the nature of the secondary ion spectra of organic materials, does not address some of the particular issues of solid polymers. This problem will be returned to in Section 4.2.5. It will be appreciated, however, that the variety of processes by which the detected ions can be formed will be reflected in spectral similarities with other forms of organic mass spectrometry: in particular, field desorption, chemical ionisation and electron impact. Of course, this statement only applies to positive secondary ions. In conventional organic mass spectrometry negative ion formation is rare and negative ion spectra are rarely recorded. As will be seen in Chapter 5 negative ion spectra are extremely important in SSIMS of polymer surfaces.

4.2.4   Cationisation

This process leading to positive secondary ions is worthy of special mention because it has been exploited to give very high analytical sensitivity. Under the normal circumstances of polymer surface analysis, cationisation involves the attachment of alkali metal ions, most often Na^+ and K^+, to sputtered neutral molecules or fragments. The process seems to be very efficient and only trace amounts of the alkali ions need be present – low level contamination from synthesis is common – such that the actual intensity of, for example, the Na^+ peak in the spectrum is unremarkable. An example of this 'natural' cationisation is given in Fig. 4.10.

Benninghoven introduced the technique of silver cationisation for enhancing quasi-molecular ion yields from involatile or thermally labile molecules of pharmacetucial or biomedical interest (Benninghoven, 1983). This involves the deposition, from solution, of a monolayer (or less) of the analyte onto a silver surface (this can be ion-etched or chemically etched foil or an evaporated film on another substrate). Yields of $[M+Ag]^+$ ions can be very high and since silver has a distinctive isotope pattern (^{107}Ag and ^{109}Ag of almost equal abundance) the cationised species are easily identified. This is not the case for alkali ion cationisation since the cations are monoisotopic or nearly so. Deliberate cationisation has a role in SSIMS of polymers, as discussed in Chapter 5.

4.2.5   Secondary ion formation from polymers

For linear/branched (as opposed to cross-linked) polymers the types of secondary ions observed can be split into four groups. The first group comprises ions representing one or more repeat units. These may be intact or minimally rearranged in order to generate a stable entity, but in either case a sequence of ions separated by the mass of the repeat unit may be observed. The second group consists of ions which are derived from the repeat unit by simple processes, e.g. the breaking of a single side-chain bond, and which are still highly characteristic of the particular polymer. The third group consists of low mass fragments which are structurally non-specific and which include atomic species and simple combinations thereof. The fourth group comprises fragments, over the whole mass range, which result from processes which lead to ions which carry almost no 'memory' of the molecular structure of the polymer. These are broad gener-

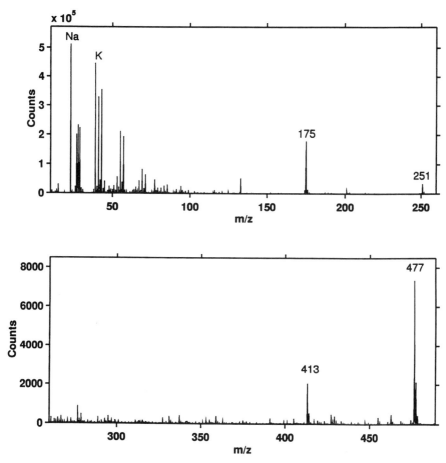

Figure 4.10   Positive ion ToF SIMS spectrum of a mixture of isodecyl diphenyl phos-
phate/di-isodecyl phenyl phosphate as a thin film on glass. The sodium cationised molecular ions, $[M+Na]^+$, at m/z 413 and 477, respectively, dominate over the $[M+H]^+$ species
(m/z 391, 455). The fragments at m/z 175 and 251 are $(PhO)P(OH)_3^+$ and $(PhO)_2P(OH)_2^+$,
respectively.

alisations only, as will be seen in Chapter 5. However, the categories of secondary
ions can be related to the processes of secondary ion formation discussed earlier.
In the region close to the primary particle impact, where the energy deposition
is greatest, the small/atomic fragments will be created. Further away, when the
energy density is reduced, fewer bonds are broken and the internal energy of desorbed species is low enough for structural rearrangements or post-desorption
fragmentation to be determined by the specific chemistry (as, for example, in
electron impact mass spectrometry). Consequently, the species retain much of
the original molecular arrangement. Finally, at the edges of the area affected by
the impact (of the order of 10 nm diameter) the large fragments are desorbed
with very little internal energy.

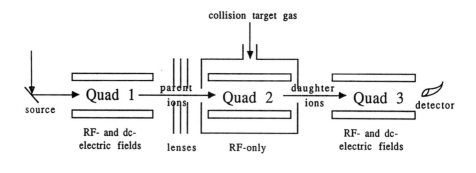

	Q1	Q3	collision cell
MS- or parent scan mode	scan	RF-only	evacuated
MS-MS or daughter scan mode	filter	scan	collision target gas admitted
neutral loss mode	scan	scan	collision target gas admitted

Figure 4.11   Schematic of the triple-quadrupole arrangement for carrying out tandem mass spectrometry experiments in SIMS (Leggett *et al.*, 1990).

Some light has been shed on secondary ion formation processes by using tandem mass spectrometry. This is a technique used in other branches of organic mass spectrometry to induce controlled fragmentation of molecular (parent) ions in order to aid analysis. Two mass spectrometers are used in tandem, separated by a collision cell. The SIMS variation on this theme employed the triple-quadrupole arrangement shown in Fig. 4.11 (Leggett, Briggs & Vickerman, 1990). There are three operational modes. If quadrupole 1 (Q1) is operated as normal (Section 4.1.4) and Q2 and Q3 only have the rf field applied, so that they transmit all ions, then the normal SIMS spectrum is obtained. In the daughter scan mode an ion of interest, selected by Q1, is passed to Q2 into which target gas molecules (e.g. argon) are leaked. Daughter ions resulting from collisionally activated decomposition of the selected (parent) ion are then analysed by Q3 which scans the range $m/z=0-M$, where M is the parent ion mass. The loss of neutrals in the course of fragmentation in the collision cell can also be measured by scanning Q1 and Q3 synchronously with Q3 offset by a fixed mass difference equal to that of the neutral under scrutiny.

Fig. 4.12 shows some illustrative results from tandem mass spectrometry. The low intensity of the daughter ion and neutral loss spectra is noteworthy; this is a direct consequence of the low transmission of the triple-quadrupole system (which severely limits the scope of these studies).

The number of polymers studied in this way is still in single figures. However, some interesting general conclusions have emerged. First, daughter ion spectra show that most of the secondary ions observed in the SIMS spectrum *could* result from fragmentation of the highest mass species observed (at least within the

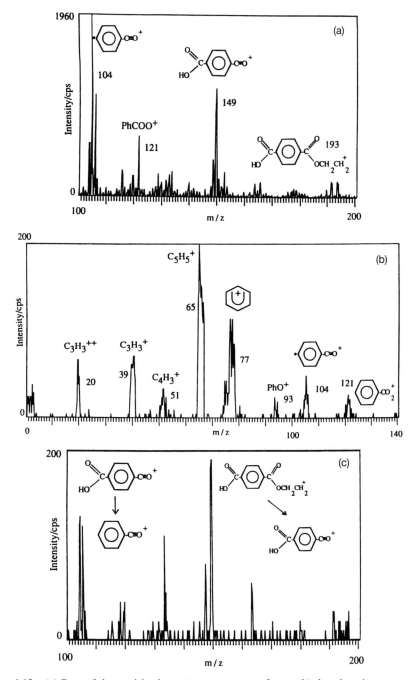

Figure 4.12 (a) Part of the positive ion SSIMS spectrum of PET; (b) daughter ion spectrum of the *m*/*z* 149 ion from PET; (c) neutral loss spectrum for 44 amu from PET over the same mass range as (a) (Leggett *et al.*, 1990).

range and sensitivity of the quadruple mass spectrometer). Second, there are some striking similarities between relative intensities of peaks in neutral-loss spectra and the SIMS spectra themselves which suggests that simple elimination processes may account for much of the observed fragmentation. Third, these similarities between the tandem mass spectometry results and the original spectra are most obvious at low collision energies (a few electron volts) which emphasises the *soft* ionisation mechanisms operating in SSIMS. Finally, though, there are prominent secondary ions which cannot be generated in such collision experiments: these may have quite complex structures not easily derived from the polymer structure. This would be consistent with mechanisms for secondary ion formation involving higher energy processes and operating in the region of direct impact by the primary particle.

Further understanding of secondary ion formation processes is now being obtained from the measurement of secondary ion energy distributions, made possible by the particular properties of the so-called TRIFT design of ToF SIMS (Section 4.1.7). As shown in Fig. 4.6, ions of a given mass are energy dispersed by the first electrostatic analyser (ESA1). A narrow slit ($100\,\mu$m) placed at the cross-over between ESA1 and ESA2 can be used to select a narrow energy band (1.5eV). By stepping the extraction voltage applied to the sample and obtaining a series of spectra, the variation in any peak intensity over a chosen energy region (the kinetic energy distribution) can be obtained (Delcorte & Bertrand, 1996a,b; Delcorte, Segda, & Bertrand, 1997).

4.2.6   SSIMS conditions and sample damage

By definition, the SIMS spectrum of a surface obtained under *static* conditions represents virgin (or undamaged) material. Since the processes leading to secondary ion formation are inherently destructive it is clear that, in the absence of rapid annealing mechanisms, the surface will be progressively damaged as particle bombardment continues and the spectrum will mirror the changes in structure. The critical parameter is therefore the primary particle dose required for acquisition of the spectrum. For convenience we consider only ion bombardment from now on; the majority of SSIMS data relate to ion bombardment and ion doses are generally easier to measure that atom doses.

A simple calculation indicates the likely restriction on this dose. Assume 10^{15} atoms in 1 cm^2 of a surface monolayer. If a single ion impact affects a region of the order of 10 nm diameter, then all the surface atoms will be affected by a dose of 10^{13} ions cm^{-2}. If a large organic molecule, or a representative segment of a polymer occupies a significant fraction of the 10 nm diameter region then its integrity will be disrupted by one impact. If the minimum requirement for SSIMS is that the *same* spectrum can be obtained from a given area of sample in successive experiments then 10^{13} ions cm^{-2} is likely to represent a damaging dose.

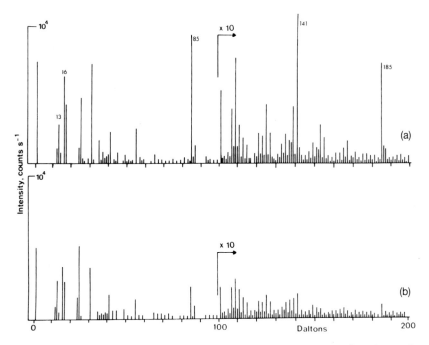

Figure 4.13 (a) Negative ion SSIMS of PMMA using 4 keV Xe^+ (0.5 nA cm^{-2}) and a total acquisition dose of 1.5×10^{12} ions cm^{-2}. (b) As (a) but following irradiation with 4 keV Xe^+ at 9 nA cm^{-2} to give a total dose of 5.5×10^{13} ions cm^{-2} (Briggs & Hearn, 1986).

That this is, in fact, the case was shown very early in the development of SSIMS for polymer analysis (Briggs & Wooton, 1982) by using XPS to examine, *in situ*, polymer surfaces irradiated under SIMS conditions; this was necessary because the changes in the SIMS spectra themselves, as a function of dose, could not be easily interpreted because of possible charging effects (see Section 4.2.7). When these problems had been overcome results such as those in Fig. 4.13 were obtained. Structurally diagnostic peaks (see Section 5.2.2) in the spectrum are virtually eliminated by doses significantly greater than 10^{13} ions cm^{-2}. Relative intensities of peaks in the spectra change markedly as shown in Fig. 4.14. In the case of acrylic polymers, of which PMMA is an archetypal example, these changes are consistent with preferential loss of the ester side-chain and the formation of cross-linked unsaturated structures which are carbon rich. Ultimately, the result would be an impure, largely amorphous, carbon. (Briggs & Hearn, 1986). Damage mechanisms under SIMS conditions are relatively unexplored, but it is clear that they are rather specific to the particular polymer concerned.

From the above studies it became accepted that 10^{13} ions cm^{-2} is the maximum dose (or threshold) for SSIMS experiments on organic materials and that doses closer to 10^{12} ions cm^{-2} were preferable to accommodate the fact that

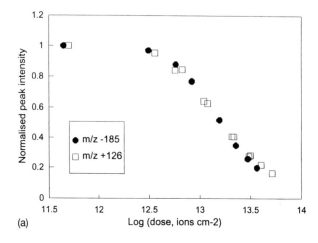

(a)

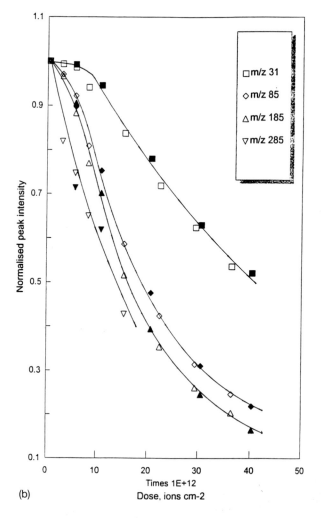

(b)

Figure 4.14(a) Variation in the absolute intensities of highly characteristic positive (m/z +126) and negative (m/z –185) molecular fragments of PMMA bombarded with 8 keV Cs+. The log scale for the primary ion dose emphasises the onset of damage (the so-called 'static SIMS limit') (Briggs & Fletcher, 1997).
(b) Variation in absolute intensities of the characteristic negative ions in the PMMA spectrum under 8 keV Cs$^+$ bombardment. The linear dose scale emphasises the differences in intensity decay rates. Open and closed symbols are for different primary ion (pulsed) current conditions (Briggs & Fletcher, 1997).

some materials are clearly more damage prone than others. As discussed in Section 4.1.4, doses in this range do allow adequate SIMS spectra to be obtained with quadrupole instruments. For instance: $1\,nA\,cm^{-2}$ primary ion beam, m/z 0–300 mass range scanned with 3000 channels and 150 ms per channel; total dose is 2.8×10^{12} ion cm^{-2} for a 7.5 min scan and under these conditions typical *maximum* peak intensities are 10^4 counts s^{-1}. Generally, spectral intensity falls rapidly with mass because of the inherently lower yields of higher mass fragments, so that by m/z 250 polymer fragment intensities are two orders of magnitude or so lower than this with counts per channel approaching noise levels. Nevertheless, such spectra are usually perfectly adequate for fingerprint recognition purposes.

With ToF SIMS instruments, spectra with superior signal:noise, and extending to m/z 10000 (if necessary) can be obtained for doses of $\sim10^{11}$ ions cm^{-2}, thus damage can be safely neglected.

Damage is a much more serious issue when imaging at high magnification. For these experiments a ToF instrument is a necessity as noted previously. A useful parameter is the minimum width (w) of analysis area which can be imaged for a material. This can be estimated from the relationship:

$$w=\frac{10^4(np)^{1/2}}{dy}\ \text{(in }\mu m) \tag{4.3}$$

where n is the minimum average number of counts per pixel in the image (required to give adequate contrast), p is the total number of pixels, d is the total primary ion dose (in ions cm^{-2}) and y is the useful ion yield, i.e. the number of analytically useful secondary ions detected per incident primary ion. Even though y can be the intensity summed over several peaks which relate to a particular chemical species, typical values fall in the range of 10^{-2}–10^{-4}. Table 4.1 gives some illustrative values of w for a total dose of 10^{13} ions cm^{-2}. One count per pixel may seem to be very small but this intensity level can provide adequate image contrast, particularly when use is made of modern image processing and display techniques. Note that for the most favourable case in Table 4.1 each pixel

Table 4.1 *The minimum width of an area which can be imaged for a total dose of 10^{13} ions cm^{-2} as a function of the number of image pixels (p), the useful ion yield (y) and the minimum average number of counts per pixel (n).*

		Image width, w in μm					
p	y	10^{-2}		10^{-3}		10^{-4}	
	n	1	5	1	5	1	5
128×128		4	9	13	29	40	90
256×256		8	18	26	57	81	181

dimension (4/128 μm) is 30 nm. Although liquid-metal ion guns can deliver spot sizes close to this when operating in continuous mode, the best current spot size for a pulsed ion gun is ~100 nm. In the general case of useful ion yields in the 10^{-3}–10^{-4} range and for somewhat less optimised ion beam diameters (say 0.2–1 μm) there is a satisfactory match between desired image dimensions ($\leqslant$100 μm), probe limited pixel resolution and counting statistics.

4.2.7   Surface potential control

The control of the surface potential of electrically insulating materials is absolutely critical to SSIMS studies. The injection of positively charged ions and the consequent emission of secondary electrons leads to positive charging; preventing this, i.e. controlling the surface potential, is usually referred to as 'charge neutralisation' or 'charge compensation'. Without such surface potential control the secondary ion energies cannot be matched to the acceptance energy range of the mass spectrometer. As charging proceeds, positive secondary ions will acquire increasing emission energies whilst negative secondary ion emission will be rapidly suppressed. The surface potential is controlled by providing a source of low energy electrons which compensate the inherent positive charging. The strategy is different for quadrupole and ToF instruments.

In quadrupole instruments extraction fields are usually relatively low so that, in principle, it is possible to direct a low energy electron beam onto the sample. In practice, however, it is usually necessary to use electrons of several hundred electron volts in order to achieve the required combination of spot size, beam steering and current density. This leads to the complication, especially with organic materials, that the electron beam can itself generate ions (a process referred to as electron stimulated ion emission, ESIE, or electron stimulated desorption, ESD). Predominantly, the cluster ions are positively charged. A satisfactory solution to this problem is to adjust the position of the electron beam between positive and negative ion collection (Briggs, 1992). For the former case the beam is outside the field of view of the collection optics so that charge compensation is by low energy secondary electrons generated from the sample edge or holder, whilst for the latter the electron beam is within the field of view and the higher electron current density drives the surface potential negative to aid negative secondary ion collection. In either case a stable surface potential can be achieved such that the secondary ion energies are within or close to the range acceptable by the quadrupole mass spectrometer (of order 10$\pm$5 eV). Final matching can be achieved by varying a bias voltage applied to the sample holder. The importance of this 'tuning' can be appreciated from Fig. 4.15. This contrasts the crude energy distributions (actually the convolution of the true energy distribution with the energy band pass of the collection optics) of atomic and cluster ions from PET. Clearly the cluster ion energy distribution is extremely

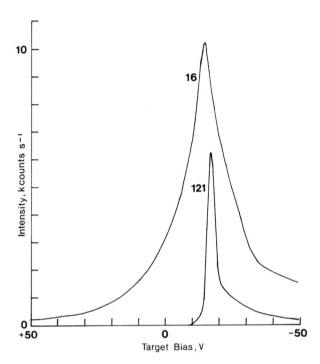

Figure 4.15 Crude energy distributions for O⁻ (*m/z* 16) and C₆H₅COO⁻ (*m/z* 121) from PET obtained by measuring peak intensity as a function of sample (target) bias (Briggs, 1992).

narrow and optimum collection of these ions requires precise control of the surface potential. Notice also how high intensity in the atomic ion peak can be obtained when there is *no* spectral contribution from the cluster ion. Different cluster ions from the same material have slightly different energy distributions (Briggs & Wootton, 1982; Delcorte & Bertrand, 1996a), consequently their relative intensities will depend on the surface potential as demonstrated by Fig. 4.16. Since these spectra take several minutes to acquire, it is obvious that surface potential drift during this period must be avoided. An ingenious development overcomes many of these problems (Gilmore & Seah, 1996). An emission plate positioned in front of the collection optics, and biased to the same voltage as the sample holder, is bombarded with high energy electrons. Low energy secondaries provide the compensating electron flood with the advantage of being created in a field-free region. The bias voltage is modulated at high frequency so that, in effect, the secondary ion energy distributions are broadened. This significantly reduces the dependence of spectral intensities on the surface potential, but at the cost of a significant loss in sensitivity.

In ToF instruments, with their high extraction voltages (several kiloelectron volts) these approaches are not possible. Instead, advantage is taken of the low duty cycle whereby the secondary ion mass analysis period is several orders of magnitude longer than the primary ion pulse duration. The pulsed electron beam method (Hagenhoff *et al.*, 1988) is shown schematically in Fig. 4.17. After a primary ion pulse the extraction field is reduced to zero volts and a long pulse

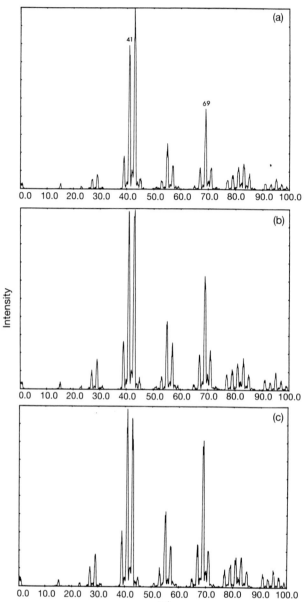

Figure 4.16 Positive ion spectra in the range *m/z* 0–100 from an oxidised polyalkene film surface obtained after optimisation of the peaks at (a) *m/z* 41, (b) *m/z* 69 and (c) *m/z* 109 (not shown in the spectrum). Applied target bias was the only variable to be altered: (a) 5.0 V, (b) 5.9 V, (c) 6.75 V (Briggs, 1992).

of low energy electrons is directed at the sample. After this pulse a short delay before raising the extraction field for the next primary ion pulse avoids the possibility of extracting any ions arising from electron induced emission into the spectrometer. In principle, it should be possible to use very low energy electrons to give charge compensation. In practice, and particularly in negative ion mode, it is often necessary to increase the electron energy. This can be achieved either at source or by biasing the sample during the electron pulse. Additionally,

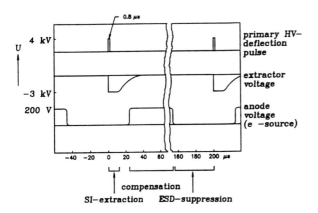

Figure 4.17 Pulse-time diagram of primary ions (upper), extractor voltage (middle) and charge compensating electrons (lower) as used for ToF SIMS of insulating materials (Hagenhoff *et al.*, 1988).

placing a metal grid or mesh over the sample and rastering the primary beam within a 'hole' helps to maintain the surface at the appropriate potential. Spectral reproducibility for insulating samples is achieved rather more easily than is the case for quadrupole instruments; however, the achievement of the highest mass resolution available from modern ToF instruments is critically dependent on the charge compensation strategy.

Charge compensation and associated ESIE problems make high spatial resolution imaging (i.e. using a liquid-metal ion source) extremely difficult with quadrupole instruments. Even with ToF instruments imaging of negative molecular fragment ions can be problematical. Complicated ion beam raster patterns, rather than simple TV (line-by-line) rastering, avoid charge localisation and are very beneficial in this respect.

The length of this section underscores the importance of charge compensation. Although successful strategies have evolved, much remains to be understood about the phenomena involved and their interplay with instrument design. Not surprisingly, the literature does not reflect the 'art' which is brought to bear in successful experimentation, some of which has been alluded to here.

4.2.8 Mass calibration and accurate mass measurement in ToF SIMS

The flight time for a singly charged ion in a linear ToF mass spectrometer is given in equation (4.1), where the total kinetic energy, $E_k = E_e + eU_0$, is the sum of the initial energy and the energy gained in the extraction field. The mass scale cannot be calibrated absolutely because small changes from sample to sample, such as the exact position of the surface with respect to the extraction plate or the exact surface potential, will critically affect the situation. Therefore, calibration is carried out *in situ*, using peaks representing secondary ions of known composition (and hence of known exact mass). This is achieved by means of the equation:

$$m = at^2 + b \qquad (4.4)$$

which derives from equation (4.1). The constants a and b can be calculated simply from two simultaneous equations in which t is the measured flight time for the chosen calibrant ion (mass m). Greater accuracy is achieved if more than two peaks are used and the constants are calculated by least squares fitting. Although the equivalent of equation (4.1) is more complicated for a non-linear ToF mass spectrometer, equation (4.4) remains valid.

Mass accuracy is defined in parts per million (ppm) as:

$$A = \left| \frac{m_{meas} - m_{exact}}{m_{exact}} \right| \times 10^6 \qquad (4.5)$$

where m_{meas} and m_{exact} are the measured and exact masses respectively. Factors which affect the accuracy of mass measurement have been specifically investigated for polymer systems (Reichlemaier et al., 1994). These include: instrumental resolution, calibration strategy, counting statistics, close mass interferences and the effect of metastables. This work showed that with current reflectron ToF designs a mass accuracy of $\sim$20 ppm should be routinely achievable and that more careful measurements, requiring spectral averaging, could improve this to $\sim$10 ppm. This is the kind of mass accuracy which is routinely achieved in gas phase organic mass spectrometry and used to identify ions on the basis of empirical formulae assignments.

Another factor, which has received less attention, is that of peak shape. Fig. 4.18 gives a fairly dramatic example of the difference in peak shape which can be encountered. Clearly, there can be no effects of differences in counting statistics or mass resolution in this case. The difference is almost certainly due to differences in the energy distributions of the secondary ions, since atomic ions from an elemental target (e.g. silicon), which are known to have very broad energy distributions, show the same peak tailing. Such differences will affect the mass calibration, particularly if the peak centroid is taken as the measured mass. Mass accuracy is improved if only peaks of similar shape to the 'unknown' are included in the calibration set.

4.2.9   The SSIMS sampling depth

This important parameter is particularly difficult to measure experimentally, and for polymer systems there is a paucity of data. However, it is very clear from many practical examples that the sampling depth of SSIMS is significantly lower than that for XPS, under typical operating conditions (i.e. take-off angles

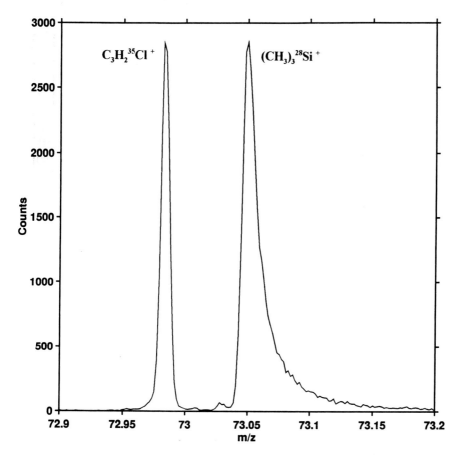

Figure 4.18 High mass resolution ($\Delta m/m \sim 10\,000$) ToF SIMS spectrum at m/z 73 of a PVC sample containing a trace of PDMS. The two peaks are representative molecular fragments from each polymer. Note the relative breadth and high mass tail of the PDMS peak.

of $>45°$, with respect to the surface). Hearn *et al.* (1987) studied the relationship between quantified XPS and SSIMS data for a series of fluorinated PUs which exhibited variable surface segregation (see also Section 6.3). Plots of the F^-/CH^- (*m/z* 19/13) ratio from negative ion SSIMS against the F/C atomic ratio from XPS gave a much better (linear) correlation with XPS data taken at a take-off angle of 10° than those taken at 90°. From this it was concluded that the SSIMS sampling depth must be of the order of 1 nm. Lub & van Veltzen (1989) came to a similar conclusion from studying the reaction of propionyl chloride with the surface $-OH$ groups of poly(hydroxyethyl methacrylate) and comparing the intensity of the propionoxyethyl side-chain fragments with those from ethyl propionate.

Delcorte *et al.* (1996) have studied the attenuation of substrate secondary ions by thin polymer layers. For Si^+ the mean emission depth was again found to be of the order of 1 nm, whilst for $SiOH^+$ and SiO_3H^+ this was reduced by factors of 2 and 3, relative to Si^+, respectively. This is in keeping with the general belief that the emission of large molecular fragments is mainly from the upper monolayer. Further evidence for the greater emission depth of atomic ions is discussed in Section 6.5.

Chapter 5

Information from SSIMS

There are inherent differences between the spectral features relating to polymers *per se* and the (usually) lower molecular weight species, such as contaminants and additives, which are often detected on polymer surfaces. It is, therefore, convenient to discuss these separately. Cationisation, deliberately induced or otherwise, leads to major variations in spectra from either class of material which are further discussed in Section 5.4. The identification of the atomic species from which either polymers or 'small' molecules are composed is, however, fairly straightforward.

5.1   Elemental identification

Most of the important elements in polymer surface analysis can be identified directly from peaks at the appropriate atomic masses. For high mass resolution instruments (i.e. $m/\Delta m > 3000$) elemental peaks are resolved from interfering organic cluster ions and identification is either by exact mass measurement or by detection of the expected isotopic abundancies. This is particularly important for the detection of low concentration metallic species which give positive ion peaks of lower intensity than the interfering organic fragment ions from either the polymer or other organic species at the surface. Fig. 5.1 illustrates the detection of zinc at the surface of a poly(styrene) sample; this would not be possible with a low mass resolution instrument. The great majority of metals give exclusively positive ions under SSIMS conditions. The most electropositive elements, i.e. the alkali and alkali earth metals, give particularly high positive ion yields. Conversely, the electronegative elements, i.e. oxygen, fluorine, chlorine, bromine and iodine, give intense negative ion signals. Hydrogen gives both H^+ and H^- in high yield.

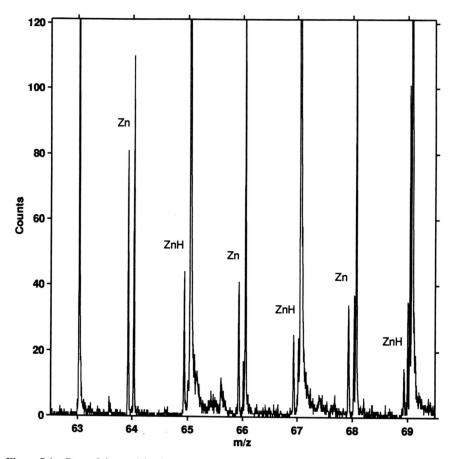

Figure 5.1 Part of the positive ion ToF SIMS spectrum from a poly(styrene) film sample. The most intense peaks (mostly expanded off scale) are $C_xH_y^+$ fragments from the polymer (see Fig. 5.3). The presence of zinc is revealed by the mass resolved peaks at below unit mass representing the isotopes at nominal m/z 64 (47%), 66 (28%) and 68 (19%) together with an equivalent pattern for ZnH.

Carbon is detected as C^-, although the accompanying CH^- is always more intense. Nitrogen is not detected via an elemental peak. If the concentration is high enough (particularly if a low resolution instrument is used) the CN^- and/or CNO^- peaks are readily observed. Depending on the oxidation state, sulphur can be observed through S^-/SH^-, SO^-, SO_2^-, SO_3^-/SO_3H^-. Phosphorus is usually identified through PO_2^- and PO_3^-.

5.2   Polymer spectra

This section deals with the spectra from bulk polymer surfaces. The relationship between the observed fragment ions and the polymer structure is a strong one, especially when heteroatom functionality is involved. Prominent ions frequently result from simple cleavage of the polymer backbone, such that the monomer chemistry is reflected, and/or pendant groups. Our understanding of polymer spectra has largely been achieved by the comparison of spectra from closely related materials having systematic variations in structure (e.g. methacrylate polymers differing only in the nature of the alkyl group in the pendant ester function). It is worth mentioning that most of this work derives from the use of quadrupole mass spectrometer instruments, prior to the availability of high mass resolution ToF systems. Consequently, much useful spectral detail may be hidden by the incorporation into one peak of components representing ions with different atomic compositions, but having the same nominal mass. Examples of this will be discussed later in the chapter.

5.2.1 Hydrocarbon polymers

The positive ion spectra of alphatic polymers are composed of $C_nH_m^+$ clusters, with $m=n,...,2n+1$. The relative intensities of peaks within each cluster and between clusters vary significantly as shown in Fig. 5.2. These spectra have been analysed in detail (Briggs, 1990a), but the following points are noteworthy. Methyl-substituted alkyl ions which are resonance stabilised probably account for the enhanced intensity of m/z 69 in the poly(propylene) spectrum and of m/z 83, 97 in the poly(isobutylene) spectrum, i.e.

m/z 69 m/z 83 m/z 97

Unsaturation leads to a greater weighting in the C_n clusters of the ions containing less hydrogen, relative to saturated polymers.

The positive ion spectra of aromatic hydrocarbon polymers are exemplified by that of poly(styrene) shown in Fig. 5.3. Compared to the spectra in Fig. 5.2 the $C_nH_m^+$ clusters extend to much higher mass ($n>24$) and have a much lower m/n as expected from the higher overall unsaturation. Most of the prominent peaks are believed to be due to cyclic and polycyclic aromatic ions (Fig. 5.4). Alkyl-substituted poly(styrenes) give similar, but distinguishable, fingerprint spectra.

The negative ion spectra from hydrocarbon polymers are very simple. Fig. 5.5 gives the spectrum from poly(styrene). Above the C_6 cluster (m/z 72–75) the spectrum is very weak. Aliphatic polymers only produce clusters up to C_4. It has been shown that the C^-/CH_2^- peak intensity ratio is sensitive to the degree of unsaturation, being about five times higher for aromatic polymers compared with aliphatic polymers (Briggs, 1990a; Chilkoti, Ratner & Briggs, 1992).

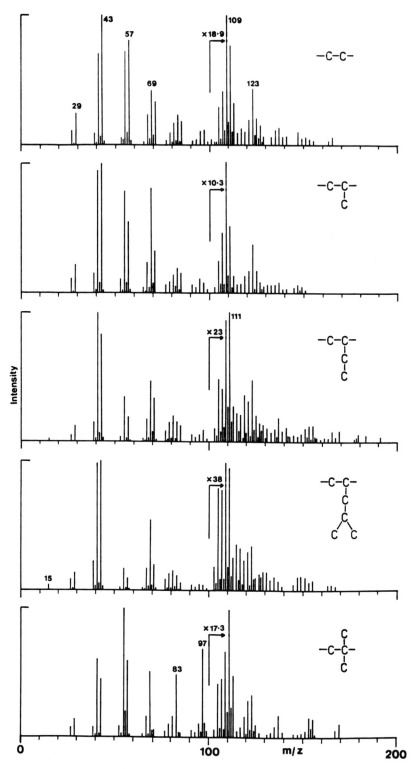

Figure 5.2 Positive ion quadrupole SSIMS spectra from the hydrocarbon polymers (top to bottom): poly(ethylene), poly(propylene), poly(l-butene), poly(4-methyl pentene-1) and poly(isobutylene) (Briggs, 1990a).

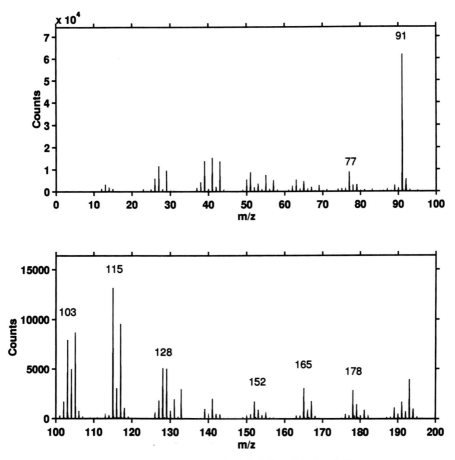

Figure 5.3 Positive ion ToF SIMS spectrum (*m/z* 0–200 only) of poly(styrene).

5.2.2 Side-chain oxygen functionalised polymers

This large class of polymers includes poly(vinyl ethers), poly(vinyl ketones), poly(vinyl carboxylates), poly(acrylates) and poly(methacrylates). In the positive ion spectra a significant contribution from the hydrocarbon ions (described above) deriving from fragmentation of the backbone is to be expected. Whether ions derived from the polymer repeat units or the functional side-chain lead to prominent peaks in the spectrum depends on their stability and hence relative yields. As a general rule positive ions of the type R−C≡O⁺ give the most intense peaks. In the negative ion spectra the problem of hydrocarbon 'chemical noise' is absent and many stable ions can be generated in which the negative charge resides on the oxygen atom, especially of the types R−C=C−O⁻ and R−C−O⁻. Families of these polymers have been studied in detail and we now summarise the salient points using representative spectra (Hearn & Briggs, 1988; Chilkoti *et al.*, 1992).

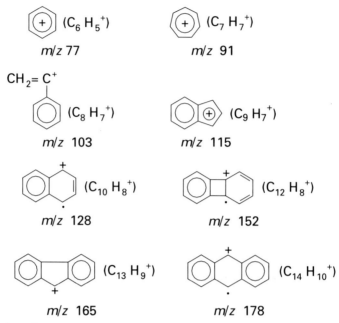

Figure 5.4 Polycyclic aromatic ion structures for the fragments highlighted in Fig. 5.3.

Fig. 5.6 compares the positive and negative ion spectra of the vinyl polymers with ether, ketone and carboxylate ester side-chains – all with ethyl as the alkyl terminating group. For the ether, the positive ion spectrum gives monomeric fragments $[M–H]^+$, $[M]^+$ and $[M+H]^+$ (*m/z* 71–73) and dimeric fragments $[2M+11]$ (*m/z* 155) with probable structures:

CH$_2$=C
   ‖
   O$^+$
   |
   C$_2$H$_5$

m/z 71

CH — CH — CH — C≡CH
‖ |
O$^+$ O
| |
C$_2$H$_5$ C$_2$H$_5$

m/z 155

Other fragments are related to these by addition or loss of backbone carbons. The negative ion fragments are essentially $C_2H_5O^-$ and $C_2H_3O^-$ (*m/z* 45, 43). C_2HO^- (*m/z* 41) is a ubiquitous low mass fragment from oxygen containing polymers.

For the ketone, by contrast, only a weak $[M+H]^+$ ion (*m/z* 85) is observed and $C_2H_5CO^+$ (*m/z* 57) dominates. This ion also dominates the ester spectrum which contains no hint of a monomer ion. Note, however, the very different probabilities of $C_2H_5^+$ (*m/z* 29) formation for these two polymers. The negative ion spectrum of the ester contains the intense carboxylate anion C_2H_5–COO^- (*m/z* 73) whilst the ketone spectrum does give repeat unit fragments, e.g.

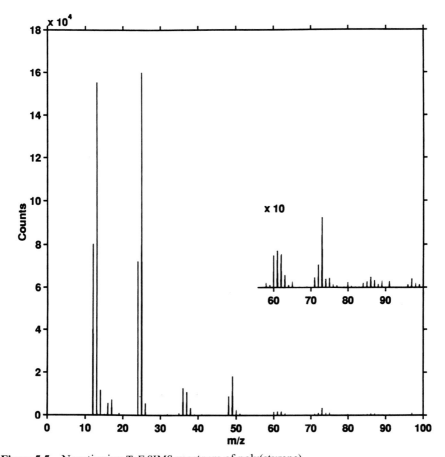

Figure 5.5 Negative ion ToF SIMS spectrum of poly(styrene).

CH — CH₂ — C = CH₂   CH₂ = C
‖ | ‖
C — O⁻   C = O   C — O⁻
| | |
C₂H₅   C₂H₅   C₂H₅

[2M–H]⁻   m/z 167   [M–H]⁻   m/z 83

with the more intense ions at m/z 155 and 71 having lost CH from these.

Fig. 5.7 shows the positive ion spectrum from PMMA. Compared with the previous spectra, the positive ion spectrum is influenced to a far greater extent by hydrocarbon fragments. However, at high resolution many peaks are seen to be split, revealing components due to oxygen-containing fragments as illustrated in Fig. 5.8. The situation is the same for all acrylic polymers. At higher mass, say m/z>100, the hydrocarbon clusters derive from backbone fragmentation and rearrangement. At lower mass they are largely due to fragmentation

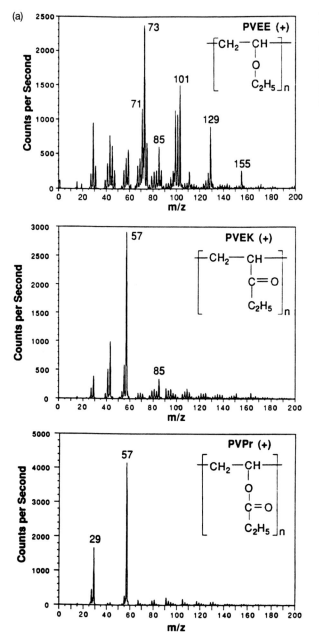

Figure 5.6(a) Positive ion quadrupole SSIMS spectra of poly(vinyl ethyl ether) (PVEE), poly(vinyl ethyl ketone) (PVEK) and poly(vinyl propionate) (PVPr).

of the ester alkyl group – often providing a specific fingerprint pattern of intensities.

The negative ion spectra of these polymers are far more informative, and that from PMMA is given in Fig. 5.9. Repeated sequences of peaks separated by 100 mass units, equal to the monomer mass, indicate multiple repeat unit fragments.

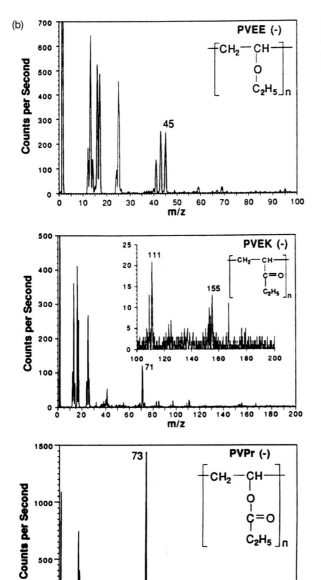

Figure 5.6(b) Negative ion quadrupole SSIMS spectra of PVEE, PVEK and PVPr (Chilkoti *et al.*, 1992).

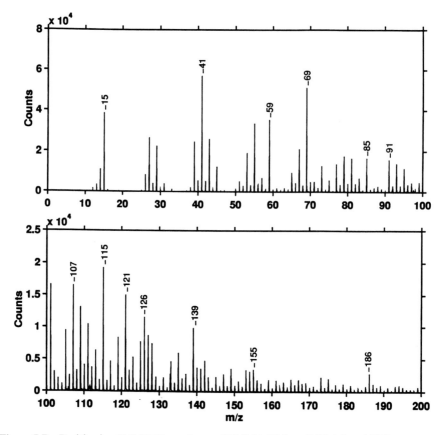

Figure 5.7 Positive ion ToF SIMS spectrum of PMMA (Briggs & Fletcher, 1997).

Negative ions at m/z 85, 109, 125 and 139 are diagnostic for all methacrylates; the structures ascribed are:

m/z 85

m/z 109

m/z 125

m/z 139

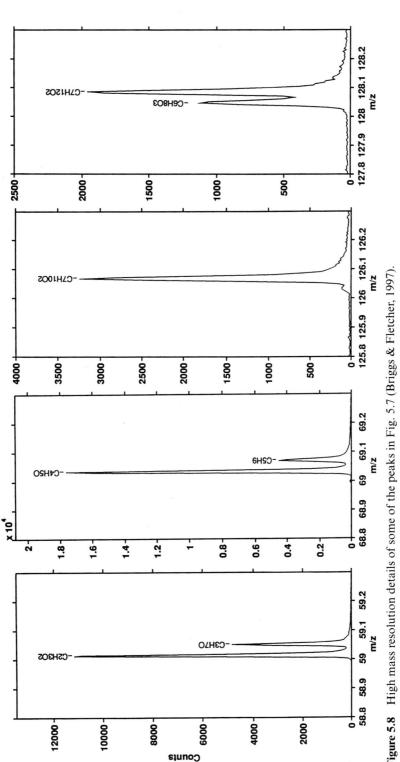

Figure 5.8 High mass resolution details of some of the peaks in Fig. 5.7 (Briggs & Fletcher, 1997).

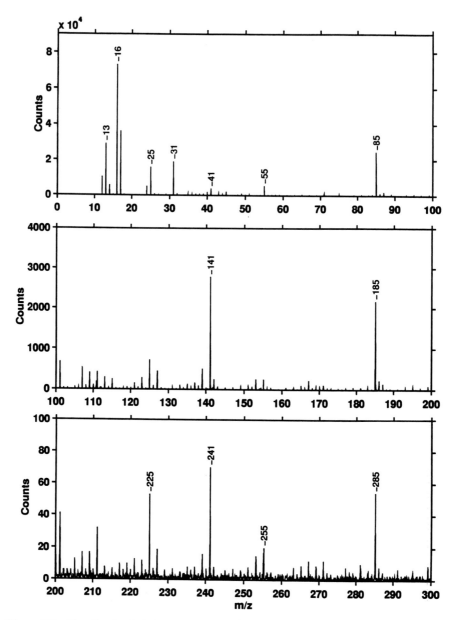

Figure 5.9   Negative ion ToF SIMS spectrum of PMMA (Briggs & Fletcher, 1997).

Ions which retain the ester alkyl group (R) are polymer specific. Some of the larger fragments have the structures:

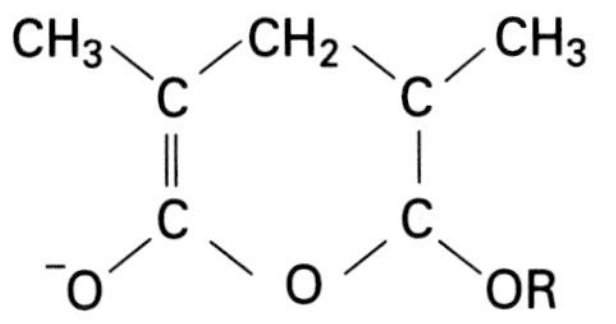

[M+55] 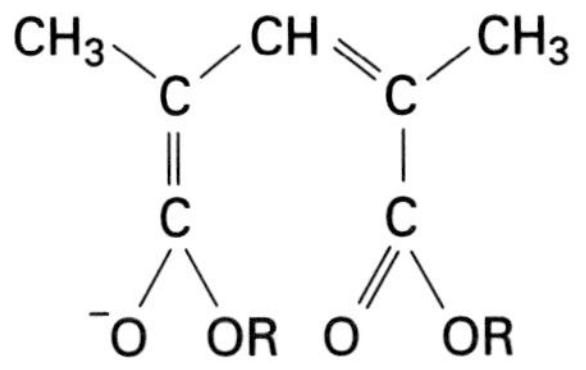 [2M–15]

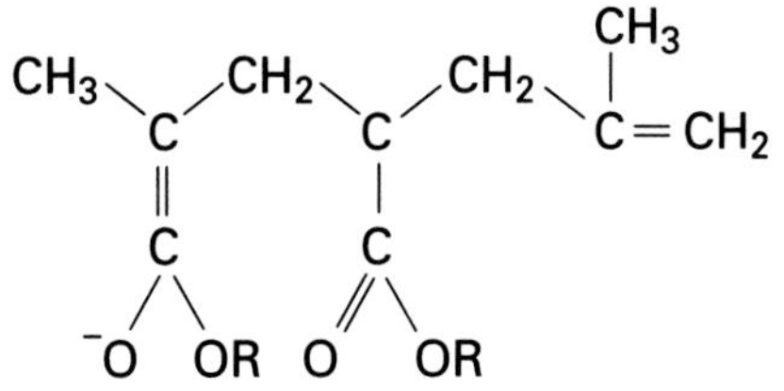

[2M+41]

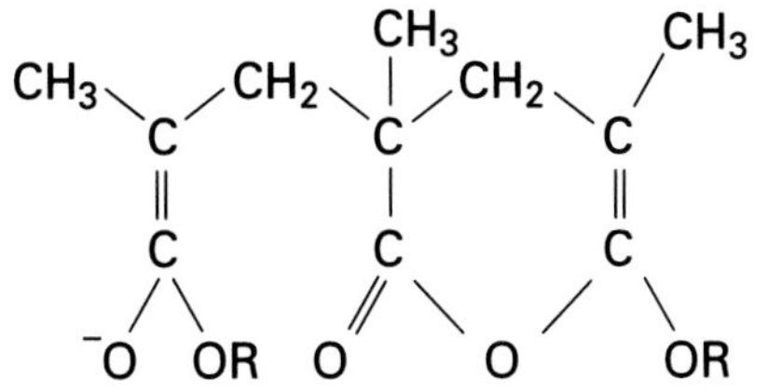

[2M+55] 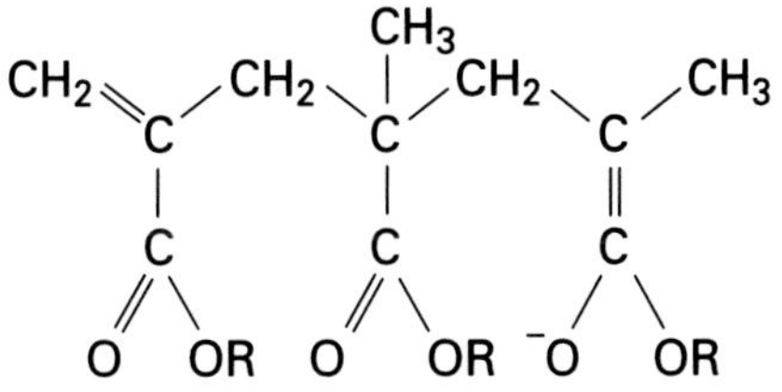 [3M–15]

In the PMMA spectrum these ions are at m/z 155, 185, 241, 255 and 285, respectively, since M=100 and R=15 (CH_3).

Although there are similarities between the fragmentation of poly(acrylates) and poly(methacrylates), the abstractable backbone hydrogen in the former leads to different fragmentation routes. Characteristic negative ions for polyacrylates are:

$$CH\equiv C-C\equiv C-O^-$$
m/z 65

$$CH_2=CH-C\equiv C-O^-$$
m/z 67

$$CH_2=CH-C-O^-$$
$$\|$$
$$O$$
m/z 71

$$CH_3-CH=CH-C\equiv C-O^-$$
m/z 81

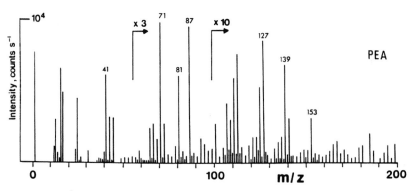

Figure 5.10   Negative ion quadrupole SSIMS spectrum of PEA (Briggs & Hearn, 1988).

whilst polymer specific ions retaining the ester group are:

$$^-CH_2-\underset{\underset{O}{\|}}{C}-OR \qquad CH_2=\underset{\underset{COOR}{|}}{C}-CH_2^- \qquad CH_2=\underset{\underset{COOR}{|}}{C}-CH_2-CH_2^-$$

$$\underset{\underset{COOR}{|}}{CH}=CH-C\equiv C-O^- \qquad CH_3-\underset{\underset{COOR}{|}}{C}=CH-C\equiv C-O^-$$

For poly(ethyl acrylate) (PEA), Fig. 5.10, these appear at m/z 87, 113, 127, 139 and 153, respectively. Multiple repeat unit fragments are less stable than for the methacrylate analogue.

5.2.3   Main-chain oxygen functionalised polymers

Linear polyethers give rise to H- or CH_3-capped oligomeric positive ions created by chain cleavage in the α or β position on either side of the oxygen atom. Thus for PEG, $(CH_2CH_2O)_n$, the main series has the formula $(CH_2CH_2O)_nH^+$. An equivalent series of negative ions has the formula $HO(CH_2CH_2O)_{n-1}CH_2CH_2O^-$.

In-chain aliphatic polyesters, either of the poly(lactone), $(R-COO)_n$, or poly(diacid-diol), $(R_1COOR_2COO)_n$ types give multiple repeat unit ions and fragments formed by cleavage at various points along the chain. Characteristic positive ions are $[nM+H]^+$ and $[nM-OH]^+$, with probable $R-C\equiv O^+$ structures. Characteristic negative ions are $[nM+H]^-$ and $[nM+OH]^-$, with probable $RCOO^-$ structures. PET, whilst having the aromatic diacid, conforms to this pattern, as shown by Figs. 5.11 and 5.12.

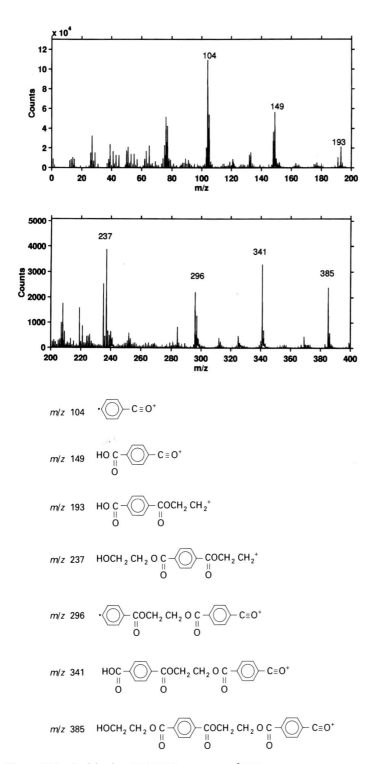

Figure 5.11 Positive ion ToF SIMS spectrum of PET.

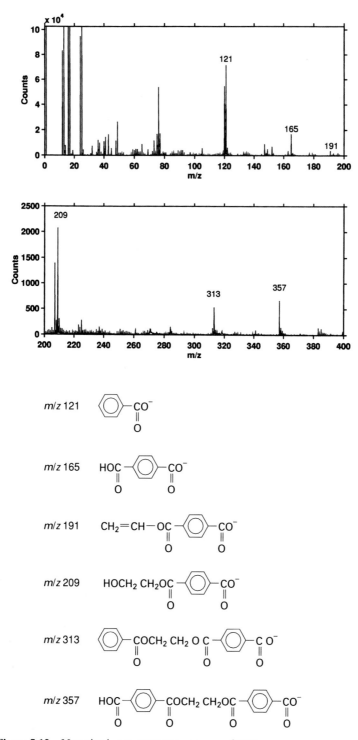

Figure 5.12   Negative ion ToF SIMS spectrum of PET.

Polycarbonate, with the repeat unit:

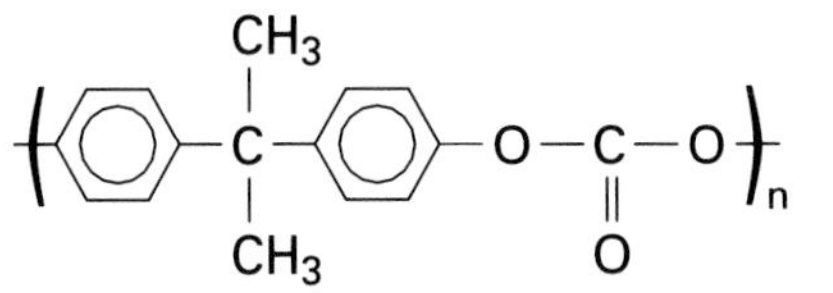

represents the class of aromatic polyethers. Stabilisation of charge on the phe-
nolic oxygen leads to easily rationalised negative ion fragments:

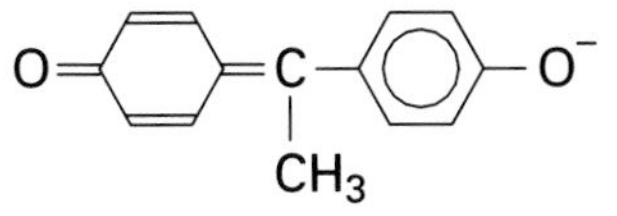

m/z 211

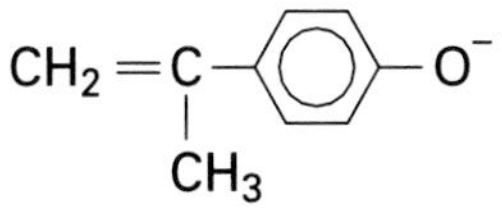

m/z 133

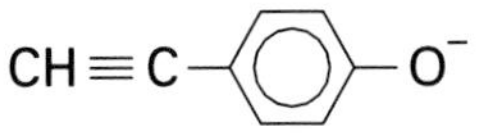

m/z 117

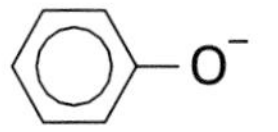

m/z 93

5.2.4   Nitrogen-containing polymers

The presence of nitrogen in a polymer is manifested in both positive and negative ion spectra by a much higher than normal incidence of even-mass peaks (due
to fragments with an odd number of nitrogen atoms) and in the negative ion
spectrum by a higher than normal intensity at m/z 26 (CN^-) and at m/z 42
(NCO^-) if the N—C—O linkage is present. These features are well illustrated in
the spectra from poly(amides) (nylons) of which poly(caprolactam) (nylon-6) is
representative, Figs. 5.13 and 5.14. Multiple repeat unit fragments $[nM+H]^+$ are
seen up to $n=9$. A characteristic of poly(lactams) is the pair of peaks $[2M+H]^+$
and $[2M-OH]^+$. In this and other respects there are similarities between
poly(amide) (both lactam and diacid-diamine types) positive ion fragments and
in-chain polyester fragments. This is not surprising since the analogous units are
R_1CONHR_2 and R_1COOR_2, respectively, and the dominant fragments are, in
each case, of the type $R-C\equiv O^+$. However, there is no stable equivalent of the
$RCOO^-$ ion from nylons and so the negative ion spectra are lacking in molecular fragments (Briggs, 1987).

Common poly(urethanes) have the general structure:

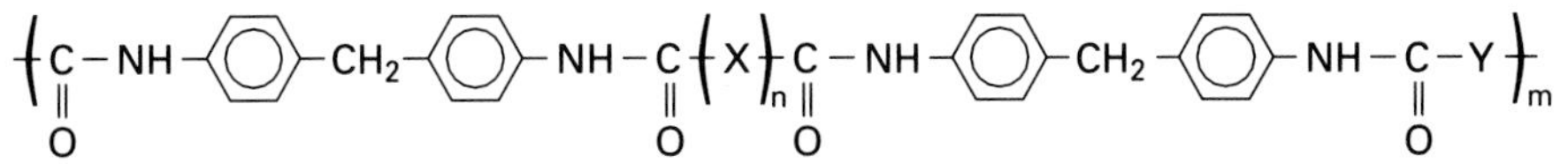

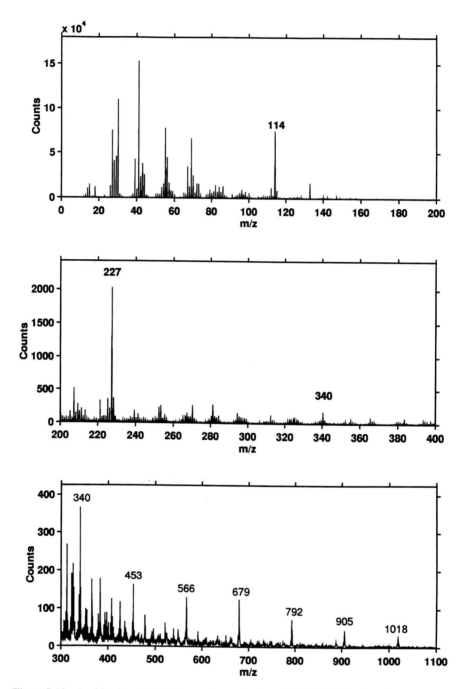

Figure 5.13 Positive ion ToF SIMS spectrum of nylon 6. The highlighted peaks represent the multiple repeat unit fragments $[n\text{M}+\text{H}]^+$ from $n=1$ (m/z 114) to $n=9$ (m/z 1018) where M is the monomer mass (113 amu).

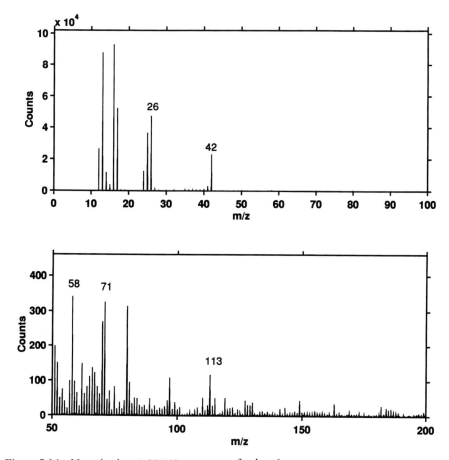

Figure 5.14 Negative ion Tof SIMS spectrum of nylon-6.

and are synthesised from 4-4′-methylenebis(phenylene isocyanate) (MDI), a poly-ether $(X)_n$ and a chain-extending diol or diamine (Y) giving a poly(etherurethane) or a poly(etherurethaneurea), respectively. The positive ion spectra are a super-position of the spectrum of the poly(ether) (Section 5.2.3) and of the rest of the polymer chain, referred to as the hard segment, derived from MDI. This produces highly characteristic fragments (Hearn, Ratner & Briggs, 1988):

m/z 233

m/z 180

m/z 132

m/z 106

5.2.5   Halogen-containing polymers

Aliphatic chlorine containing polymers give positive ion spectra which at unit mass resolution strongly resemble those from purely hydrocarbon polymers, with no obvious evidence of chlorine-containing fragments either from unusual peaks or the distinctive $3:1$ ($^{35}Cl:^{37}Cl$) isotopic abundance pattern. However, a detailed high mass resolution study of PVC, $(CH_2CHCl)_n$, has shown that the majority of peaks in the spectrum above m/z 47 (CCl^+) are split: besides the hydrocarbon ions there are components of the same nominal mass with one or two chlorine atoms (Briggs & Fletcher, 1997). This situation is likely to be general. The negative ion spectra give intense Cl^- peaks, much weaker Cl_2^- peaks and, if the chlorine concentration/structure is appropriate (e.g. chlorinated PVC), even weaker Cl_3^- peaks.

Flurone containing polymers similarly give intense F^- and weaker F_2^- negative ions. CF_3^+ (m/z 69) is usually an intense generic fragment whatever the polymer structure (CF_3 may not be present as a unit). PTFE, $(CF_2CF_2)_n$, gives fragments with the general formula $C_xF_y^\pm$ extending to high mass ($m/z > 400$). Similarly, poly(vinylidene fluoride), $(CH_2CF_2)_n$, gives fragments with the general formula $C_xH_yF_z^\pm$.

5.2.6   Silicones

Silicone rubbers are an important class of polymer, but silicone oils have a special place in polymer surface analysis because they constitute the single most common form of contamination leading to adhesion problems. These are usually dimethyl silioxane, $(Si(CH_3)_2O)_n$, polymers. The positive ion spectrum shown in Fig. 5.15 is instantly recognisable. Some of the characteristic peaks are believed to be due to linear or cyclic structures:

	n	0	1	2	3
$[(CH_3)_3 Si-(O-Si(CH_3)_2)_n]^+$	m/z	73	147	221	295

	n	0	1	2
	m/z	133	207	281

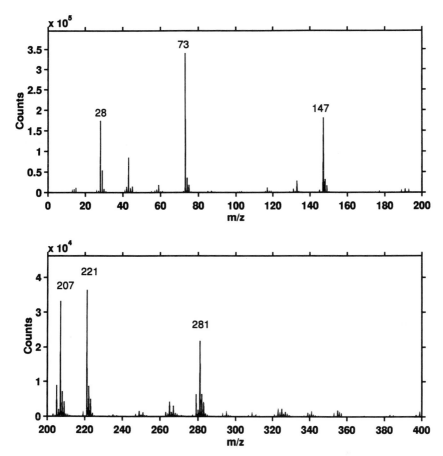

Figure 5.15 Positive ion ToF SIMS spectrum of poly(dimethylsiloxane).

Some polymers may contain large fractions of cyclic oligomers and rubbers may be cross-linked; these variations can give rise to significant variations in the relative peak intensities within the same pattern of characteristic peaks.

5.2.7 Quantification aspects

From the discussion of our understanding of the SSIMS process in Chapter 4 it will be clear that there is no immediate possibility of a 'first principles' approach to quantification from SSIMS peak intensities. However, there are many examples of the derivation of quantitative trends in composition from relative peak intensities.

Relative intensities of atomic or quasi-atomic ions show good correlations with quantitative atomic concentrations obtained by XPS, especially when restricted to compositional trends within similar materials. Thus, whilst for a wide variety of oxygen functional polymers the $O^-:CH^-$ ratio correlates with

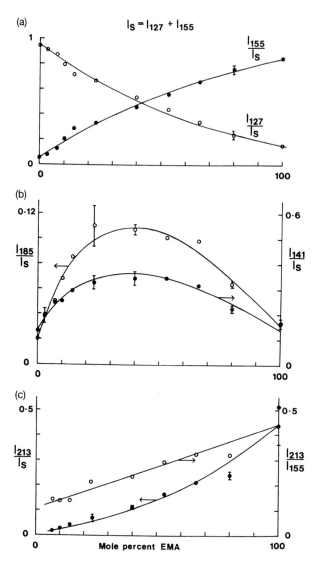

Figure 5.16 Relative intensities of characteristic ion fragment peaks from EMA–HEMA copolymers (see text) plotted as a function of bulk composition. Error bars in (a) represent the *largest* spread of values for any measurement, those in (b) and (c) represent the spread of values for all repeated measurements (Briggs & Ratner, 1988).

the XPS O:C atomic ratio, better correlations are established within each polymer class (e.g. methacrylates, vinyl carboxylates) (Chilkoti *et al.*, 1992). Similarly, for a series of fluorinated poly(etherurethanes) the $F^- : CH^-$ correlates with the XPS F:C atomic ratio (Hearn *et al.*, 1987).

   In situations where surface segregation effects are absent, so that the surface composition sampled by SSIMS is the same as the bulk composition, trends in the relative intensities of peaks representing discrete components of copolymer systems give close to linear correlations with bulk composition. This has been demonstrated for a wide variety of copolymer types. One system studied in some statistical detail, illustrated in Fig. 5.16, is that of random copolymers of ethyl-methacrylate (EMA) and hydroxyethyl methacrylate (HEMA). Following the discussion in Section 5.2.2, the negative ion spectra furnish fragments characteristic of each monomer and also of linked monomers. The following ions are used in the analysis:

m/z 213   *m/z* 185   *m/z* 155

m/z 141   *m/z* 127

m/z 155 and 213 represent monomer and dimer fragments of PEMA. *m/z* 127 is characteristic of PHEMA (loss of $^{+}CH_2CH_2OH$ from the side-chain) and thus *m/z* 185 represents a HEMA–EMA link. Ethylene elimination from the EMA side-chain occurs, hence *m/z* 141 also represents a HEMA–EMA link. The intensities of these fragments are denoted I_F. Taking the ratio $I_F:I_S$, where $I_S=I_{127}+I_{155}$, is a simple way of normalising peak intensities since, if matrix effects are small, I_S should represent the total composition. This is seen to be reasonably true from Fig. 5.16(a) (that the curves do not begin or end at 0 or 1 reflects the small contributions to each intensity from component(s) of the other monomer). The HEMA–EMA fragments show a maximum intensity at close to 50 mole %, as expected for a random copolymer (Fig. 5.16(b)), and the intensity of EMA–EMA fragments increases rapidly with the EMA content (Fig. 5.16(c)). The ratio I_{213}/I_{155} represents the ratio of EMA diads to singles. For a random AB copolymer for which diad (AA) signals come from sequences B–$(A)_n$–B where $n>2$, a statistical analysis predicts the ratio of diads to singles (isolated A units) will be proportional to the mole fraction of A, as seen in Fig. 5.16(c). These studies (Briggs & Ratner, 1988) utilised a quadrupole instrument. With a high mass resolution ToF instrument, allowing access to larger fragments (more repeat units) and the resolution of mass interferences, sequencing studies are potentially possible (Brinen *et al.*, 1993).

5.2.8   End-group effects

Even for quite high molecular weight ($>10^4$) polymers end-group fragment ions can be prominent in spectra from bulk surfaces. This can be for one or more of the following reasons: end-group fragments are more easily formed, relative to in-chain fragments, because only one bond needs to be broken; end-group fragments may be very stable and give high ion yields; end-groups may segregate to give an enriched surface concentration.

A good example is afforded by poly(carbonate). In the negative ion spectrum, in addition to the characteristics peaks described in Section 5.2.3, a peak of significant intensity can be identified as the end-group fragment:

$$R-\!\!\bigcirc\!\!-\!O^- \qquad R = {}^tC_4\,H_9 \qquad m/z\ 149$$

$$R = {}^iC_8\,H_{17} \qquad m/z\ 205$$

where the t-butyl is most commonly encountered. Reihs et al. have shown that the ratio of repeat unit to end-group fragments (I_1/I_e) is simply proportional to the average molecular weight as derived from gel permeation chromatography. This result is predicted from a kinetic fragmentation model which assumes an overall first order rate constant for fragment formation (Reihs et al., 1995). For a known average molecular weight polymer, deviations from the expected I_1/I_e ratio can then be used to infer surface segregation of end-groups. Similarly, a correlation has been established between such a ratio and the average number of monomer units in the Krytox family of perfluoropolyethers, repeat unit $(CF(CF_3)-CF_2-O)_n$ (Fowler et al., 1990).

A common end-group for PMMA is $-H$. Fig. 5.17 shows part of the negative ion spectra from high molecular weight (a) and low molecular weight (c) polymers. The additional peaks at m/z 187 and 201 in spectrum (c) are believed to be due to the two different end-groups and this is reinforced by the appearance of peaks at m/z 87 and 101 at lower mass (chain end fragment minus one repeat unit of m/z 100). When PMMA is treated briefly with an ammonia plasma (b) the m/z 187 (and 87) peak is introduced suggesting that chain scission results, with the production of mainly $-H$ end-groups. Rinsing in ethanol results in reversion to spectrum (a) showing that the lower molecular weight material is easily washed off the unaffected bulk polymer (Lub & van Vroonen, 1989).

Chain end effects in PEG spectra are considered in Section 5.2.10.

A particular case of intense spectral contribution from end-group-derived secondary ions is provided by charge-stabilised colloidal polymers (latex polymers in which the initiator provides the charged end group). Davies et al. have studied poly(n-butyl methacrylate) (PBMA) latices prepared by persulphate initiation in the presence of surfactant or stabiliser. Even at low initiator

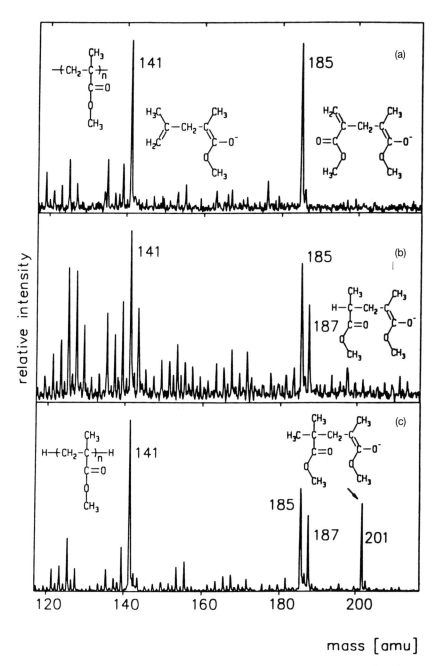

Figure 5.17 Negative ion SSIMS spectra and structures of the ions formed in the *m/z* range 120–210 of the surfaces of (a) high molecular weight PMMA, (b) high molecular weight PMMA after 5 min ammonia plasma treatment, (c) low molecular weight , (Lub & van Vroonen, 1989).

concentrations the negative ion SSIMS spectra show intense peaks due to sul-
phate end-groups. Thus in Fig. 5.18 the expected polymer peaks at m/z 55, 71,
73, 85, 109, 129, 139 and 183 (see Section 5.2.2) are matched in intensity by sul-
phate-derived ions (m/z 64, 80, 96, 97) and proposed end-group ions (m/z 119,
135, 135, 137, 151, 153). The structure:

$$^{-}O - \overset{\overset{\displaystyle O}{\|}}{\underset{\underset{\displaystyle O}{\|}}{S}} - OCH_2\,CH_2 - CH = O$$

m/z 153

was tentatively assigned to the latter. Freeze dried latex dissolved in chloroform
and cast into a film showed little evidence of these peaks, even for polymers with
over ten times as much initiator present. This shows that the charged end-groups
are indeed orientated out from the particle surface, as expected from their colloid
stabilising effect (Davies *et al.*, 1993).

5.2.9   Application of chemometrics to SSIMS data

Chemometrics is a general term for the mathematical/statistical treatment of
large data sets arising from chemical measurements. The aim is to reduce the
dimensionality of the data in order to extract relationships existing within the
datasets or between these and some other property (e.g. polymer cross-link
density, colour, surface energy). Clearly SSIMS data is information-rich and large
mass range/high mass resolution spectra constitute very large datasets. The
human interpreter can only handle two dimensional relationships within the
datasets easily, and three with greater difficulty. Therefore to make full use of *all*
information buried within the datasets requires reduction of the dimensionality;
various methods within the chemometrics 'family' do this in different ways and
may be applied either singly or in combination according to the nature of the
analytical challenge (Chilkoti, 1996). It is not possible to describe these here, but
the techniques of principal component analysis, multivariate statistical analysis,
hierarchical cluster analysis and neural networks are now being applied to SSIMS
data. These applications may be largely classification orientated (to discover the
particular features in spectra which correlate most strongly with a known variable, e.g. molecular weight, tacticity, cross-link density) or quantification orientated (to establish a quantitative relationship between spectral variables and, for
example, copolymer composition/degree of randomness, surface functionality,
surface coverage by a contaminant). The necessary tools – present generation
PCs and chemometrics software – are readily available but successful application
of these techniques does require some skill.

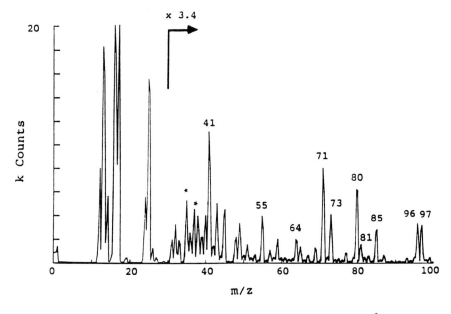

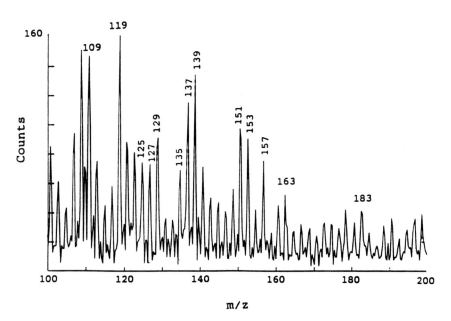

Figure 5.18 Negative ion SSIMS spectrum of a PBMA latex prepared with 14.8×10^{-5} mol dm^{-3} potassium persulphate (Davies *et al.*, 1993).

5.2.10   Use of isotope specificity

As in the early days of gas phase, electron impact mass spectrometry, replacement of ^{1}H by ^{2}D or ^{12}C by ^{13}C has been used to aid the understanding of polymer fragmentation mechanisms in SSIMS. For example, information on the facility of hydrogen transfer in the SSIMS of poly(styrene) has been obtained by comparing the spectra of all-hydrogen poly(styrene), backbone-substituted poly(d$_3$-styrene) and ring-substituted poly(d$_5$-styrene) (Chilkoti *et al.*, 1991b). Many of the ambiguities in spectral interpretation resulted from the unit mass resolution of the instruments used and these can easily be overcome by the use of current high mass resolution instruments. One example of a situation where this would not be the case is the SSIMS study by Occhiello *et al.* (1990) of the oxygen plasma treatment of poly(propylene). They used ^{16}O$_2$ and ^{18}O$_2$ for plasma treatment/post plasma exposure and *vice versa* in order to elucidate the relative importance of oxygen uptake by direct plasma modification and by reaction of plasma generated free radicals upon air exposure.

Returning to the subject of the previous section, isotopic exchange has been used to study the effect of end-group fragmentation on the positive ion spectra of PEG oligomers (Shard, Davies & Schacht, 1996). The end-groups in common PEGs are $-$OH. Low molecular weight PEGs show markedly higher intensities of the $[n\mathrm{M}+\mathrm{H}]^+$ peaks (m/z 45, 89 and 133 measured) compared with high molecular weight polymers. These fragments can arise either by single bond scission to include the end-group or by a more complicated process by which an in-chain fragment extracts hydrogen. By exchanging end-group hydrogen by deuterium to give $-$OD terminated polymers it was possible to show that end-group-derived fragments are indeed major contributors to the $[n\mathrm{M}+\mathrm{H}]^+$ ions. Similarly, the contribution of end-group-derived ions to the negative ions $[n\mathrm{M}+\mathrm{OH}]^-$ could be assessed.

5.3   Spectra from 'small' molecules

One of the major advantages of SSIMS over XPS is its ability to identify 'small' (in comparison with high molecular weight polymers) molecules on polymer surfaces. This is partly due to the greater surface sensitivity of SSIMS, which greatly weights the spectral intensity in favour of species within the uppermost monolayer, but is mainly due to the fact that these species usually give quasi-molecular ions, i.e. $[\mathrm{M}+\mathrm{H}]^+$, $[\mathrm{M}-\mathrm{H}]^-$, and characteristic fragmentation patterns. Since for most polymers the intensity in either positive or negative ion spectra falls rapidly by $m/z>250$, surface molecules which give rise to inherently high intensity quasi-molecular ions/fragments in this

region of the spectrum can easily be detected at fractional monolayer coverage. This statement needs to be qualified in respect of the type of instrument
used: although quadrupole mass spectrometers often have the mass range
necessary to detect molecular ions, they may not have the sensitivity and furthermore ion beam damage during acquisition can severely affect detection
of low surface concentration species. These problems are overcome with
current ToF instruments. Sometimes cationisation occurs naturally through
the presence of alkali (sodium, potassium) impurities to give $[M+Na]^+$ etc.
quasi-molecular ions which often have higher yields than $[M+H]^+$ (see
Section 5.4.2). Identification of surface molecules through direct interpretation of the SSIMS spectrum, in the manner of electron impact mass spectrometry is not yet possible. Electron impact mass spectra obtained under normal
low pressure conditions are dominated by odd-electron ions (molecular ion
$M^{\ddot +}$ and fragments) and the rules for their rearrangement and unimolecular
decomposition are well established (McLafferty & Turecek, 1993). By contrast, in SSIMS odd-electron ions are uncommon. This is not surprising since
they are formed in a locally high pressure region (the selvedge).
Consequently, SSIMS spectra are quite different from normal electron impact
spectra, but more like electron impact spectra obtained under high pressure
($\sim$100 Pa) conditions and/or chemical ionisation spectra in which the ions are
formed at high pressure in the presence of a gas such as methane (generating
'reagent' ions such as CH_5^+ which can react with the target molecule to
produce $(M+H)^+$ etc).

Identification of surface molecules by SSIMS, therefore, is currently heavily
reliant on pattern recognition and experience. (In the future, as SSIMS databases
expand and software 'search engines' become available, identification will
become much easier.) However, many of the molecules which appear on polymer
surfaces are members of homologous series, or are otherwise closely related
structurally, so that knowledge of the important spectral features of one molecule can often be applied generically.

The types of small molecule which find their way into polymers and which
may therefore be present as surface species were discussed in Section 1.3. The list
is a long one and discussion of the SSIMS spectra of even a fraction of the molecules would be beyond the scope of this book. Table 5.1 provides a guide to
identification of molecules frequently encountered in polymer surface analysis.
The only current collection of interpreted SSIMS spectra is *The Static SIMS
Library* (SurfaceSpectra, 1997).

Several points need to be borne in mind when using SSIMS to detect these
species. Firstly, the molecule may be volatile under analysis conditions (typically
$\sim$10^{-9} Torr). Sample cooling will often avoid this problem but, ideally, this needs
to be done prior to introduction into the vacuum system (this can be achieved
by cooling in a nitrogen gas atmosphere but requires specialised equipment and

Table 5.1 *Characteristic secondary ions and nominal masses of some molecules frequently encountered on polymer surfaces (full spectra and ion structures can be found in Briggs, Brown & Vickerman (1989) or The Static SIMS Library (SurfaceSpectra, 1997))*

Molecule	Structure	Characteristic secondary ions (m/z)	
		Positive	Negative
Poly(dimethyl-siloxane) (silicone)	CH_3 $\{Si-O\}_n$ CH_3	73,147,201, 221,281	59,60,75,119, 149,223
Poly(ethylene glycol)	$\{CH_2\,CH_2\,O\}_n$	45,89,133, 177	61,105,149
Poly(propylene glycol)	CH_3 $\{CH-CH_2\,O\}_n$	59,117,175, 233	57,75,113, 133
Stearic acid	$C_{17}H_{35}COOH$	267,285	283
Palmitic acid	$C_{15}H_{31}COOH$	239,257	255
Stearamide	$C_{17}H_{35}CONH_2$	284	42,282
Oleamide	$C_{17}H_{33}CONH_2$	282	42,280
Erucamide	$C_{21}H_{41}CONH_2$	338	42,336
Ethylene bis-stearamide	$(CH_2NHCOC_{17}H_{35})_2$	282,310	42
Di-octyl phthalate	$\overset{O}{\overset{\|}{C}}-O\,C_8H_{17}$ $C-O\,C_8H_{17}$ $\|$ O	149,167,261, 279	121
Nonyl phenol ethoxylate	$C_9H_{19}-\bigcirc-O(CH_2CH_2O)_n\,H$	247,419/435, 463/479 etc.	133,219
Glyceryl monostearate	$CH_2OCOC_{17}H_{35}$ $\|$ $CHOH$ $\|$ CH_2OH	267,341,359	71,283
Triphenyl phosphate	$(C_6H_5O)_3P=O$	77,251,327	63,79,93,249, 325

Table 5.1 (*cont.*)

Molecule	Structure	Characteristic secondary ions (*m/z*)	
		Positive	Negative
Irgafos 168 (TM)[a]	$\left(^{t}Bu-\!\!\bigcirc\!\!-O\right)_{3}P$ with ^{t}Bu	57,441,647	63,79,205, 457,473
Irganox 1010 (TM)[a]	$\left(HO-\!\!\bigcirc\!\!-CH_2CH_2COOCH_2\right)_4 C$ with ^{t}Bu groups	57,203,219, 259	59,205,231, 277
Di-octyl sulpho- succinate Sodium	$CH_2-COOC_8H_{17}$ $\mid$ $CH-COOC_8H_{17}$ $\mid$ $SO_3^-Na^+$	23,126,149, 467	64,80,81,107, 127,421
dodecyl sulphate	$C_{12}H_{25}OSO_3^-Na^+$	23,110,126, 149,165,311	80,97,119, 183,265

[a]Ciba Geigy Ltd.

careful attention to experimental procedure, especially the avoidance of surface ice formation).

Secondly, many additives used in polymer formulations are only of technical grade purity. Minor components may give rise to intensities in the spectrum which are disproportionately high relative to their bulk concentration either because they are the most surface active (and therefore 'cover up' other components) or because they have inherently higher ion yields than the main component. An example is given in Fig. 5.19.

Thirdly, the spectrum of a pure standard may differ from that of the same molecule present at a polymer surface. There are several possible causes of this effect but all of them can be classed as 'matrix effects'. Molecules within a multi-layer environment experience a different matrix (from which to be desorbed/ionised) from those in a monolayer. This leads to a likelihood that dimer (or higher *n*-mer) ions will be formed in the former case, but not the latter. This is specially noticeable if the molecule is a salt, e.g. Na^+X^- giving rise to $Na(NaX)_n^+$ and (though less likely) $X(NaX)_n^-$ clusters from a multilayer. If more than one type of molecule is present in a truly mixed surface layer (as opposed to separate surface domains) then the ionisation behaviour of each could be

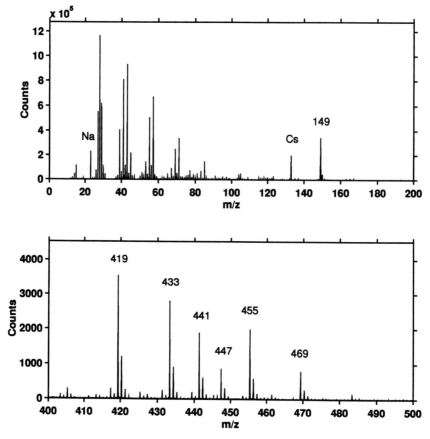

Figure 5.19   Positive ion ToF SIMS spectrum of the plastiser di-isononyl phthalate deposited as a non-uniform film on silicon (Cs$^+$ primary ions). Peaks at m/z 419 and 441 represent $[M+H]^+$ and $[M+Na]^+$ respectively. The pairs of peaks at m/z 433/455 and 447/469 are the equivalents for impurities containing one and two extra CH_2 units respectively and their combined intensity exceeds that of the principal component. m/z 149 is a common fragment of all these parent ions (protonated phthalic anhydride ion).

affected. For instance, one molecule may be a rich source of H$^+$, thereby providing a mechanism for formation of $[M+H]^+$ ions from the other molecule, M, which normally does not give high yields from the pure solid. Related to this, and considered further in the next section, is the variable potential for alkali ion cationisation. For those molecules which readily cationise, apparently very small changes in alkali ion concentration or environment can lead to major changes in spectra.

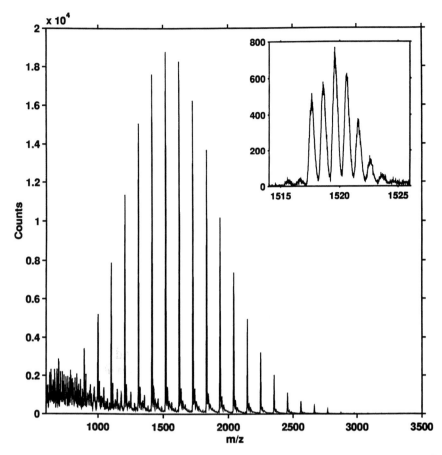

Figure 5.20   Oligomer distribution from a poly(styrene) sample with a peak average molecular weight $(\overline{M}_p)$ of 1700 $(\overline{M}_w/\overline{M}_n=1.06;$ $\overline{M}_w$ is the weight average molecular weight and $\overline{M}_n$ is the number average molecular weight) deposited on silver. The inset shows the structure of the most intense peak in the envelope $(m/z\sim1520)$ which is due to the various isotopic contributions to the ion

$$[C_4H_9\,(CH_2-CH)_{13}\,H+Ag]^+$$
$$|$$
$$C_6H_5$$

representing the intact oligomer with 13 monomer units.

5.4   Cationisation

5.4.1   Cationisation of oligomers using silver

Although this is strictly a SSIMS technique for low molecular weight polymer characterisation, it is not without relevance to true polymer surface analysis. The polymer is dissolved in a suitable solvent at a concentration of $\sim10^{-2}$M and a

1 μl aliquot deposited on about 1 cm^2 of silver foil, previously etched in nitric acid (Bletsos *et al.*, 1985). This results in a surface coverage of a monolayer or so (optimum results are obtained when the monolayer is incomplete, i.e. some silver patches are present). A typical ToF SIMS spectrum is shown in Fig. 5.20. The oligomer distribution in the low molecular weight polymer is clearly represented. Each peak in the low resolution spectrum is due to an $[M_n + Ag]^+$ species, where M_n is the mass of the intact oligomeric component, and the peak separations reveal the monomer mass. At high mass resolution the full isotope pattern for each species (involving e.g ^{107}Ag, ^{109}Ag, ^{12}C, ^{13}C, ^{1}H and ^{2}H contributions for a hydrocarbon polymer) is revealed. The accurate mass of any one of these isotopic contributions is given by:

$$nR + E_1 + E_2 + Ag = M_n + Ag$$

where R is the repeat unit (monomer) mass, E_1 and E_2 are the two end-group masses and Ag is the appropriate silver isotope mass. This allows the end-group masses to be established – as a further check the isotopic distribution can be calculated for the empirical formula and compared with the experimental distribution. Thus for the poly(styrene) used in the example in Fig. 5.20 the end-groups are C_4H_9- and $H-$ giving peaks at m/z $[58 + (n \times 104) + 107]$ for the ^{107}Ag isotope and major C/H isotopes.

The spectrum also contains two sets of lower mass ions. The pattern below the oligomers is due to cationised fragments of the intact oligomers (generally resulting from one chain cleavage) whilst at much lower mass (say m/z 0–250) the fragments typically obtained from the bulk polymer, and discussed in previous sections, appear. As the mean molecular mass increases above ~5000 the relative intensity of the cationised fragments decreases rapidly for most polymers and intact oligomers are difficult to generate beyond ~m/z 10000. This is in keeping with the dimensions of the polymer chain in the sub-monolayer coverage regime on a metal and the area of sample affected by a primary ion impact (approximate diameter 10nm), i.e. it is unlikely that for a spatially extended chain of m/z 10000 one impact will be sufficient to desorb it intact and the more likely result is desorption in a second event following chain scission by a previous impact.

This technique is relevant to polymer surface analysis in two ways. Firstly, low molecular weight species can be selectively removed from the surface by brief exposure to solvent and then redeposited onto silver. This requires some trial and error to achieve the optimum silver coverage since the extract concentration may be difficult to adjust accurately. However, the technique is incredibly sensitive and requires only minute amounts of extracted material (1 μl of a solution of an M=300 molecule at 10^{-3} mol l^{-1} concentration contains about 30ng). An example is given in Section 6.4

Secondly, the actual polymer surface can be patterned with silver by evaporating a thin layer of the metal through a grid used as a mask (Linton *et al.*, 1993). Species on the surface migrate from the uncovered surface (or through grain boundaries) onto the high energy silver patches. Thermodynamics would predict the coverage to stabilise at the monolayer level. Analysis of the silver patches then gives cationised spectra of the migrating species, often with much higher molecular ion yields than in the absence of cationisation. Although the technique implies the requirement for spatial resolution/imaging, this is not actually needed if the area ratio of silver/bare polymer is high enough (say>10).

5.4.2 Natural cationisation by alkali ions

Many molecules which give $[M+H]^+$ ions will, in the presence of alkali metal cations give $[M+A]^+$ ions (A-alkali metal) instead of, or in addition to, the protonated analogue (Fig. 5.19). Cationisation mechanisms and efficiencies are still poorly understood. $[M+Na]^+$ cations are the most commonly observed, probably because sodium is the dominant alkali metal impurity in most instances, although $[M+K]^+$ may also be observed at the same time. Surface molecules which are particularly susceptible to cationisation in this way are those containing PEG chains (these include many surfactants and antistatic agents). An interesting example is the behaviour of the random copolymers based on methyl methacrylate and poly(ethylene glycol) methacrylate (PEGMA) (Briggs & Davies, 1997). In the example illustrated in Fig. 5.21 (shown overleaf) the macromolecular PEG side-chain has an average molecular weight of ~ 1000 and fragments of this chain cationised by both Na^+ and K^+ are observed. Note how the relative intensity of the two types changes with increasing mass in this example.

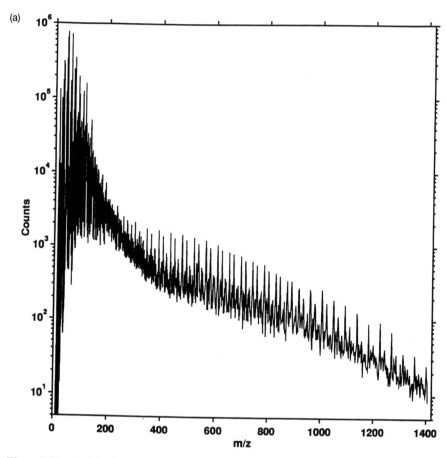

(a)

Figure 5.21   Positive ion ToF SIMS spectrum of a random copolymer of methyl methacrylate and PEGMA (80 wt% PEGMA) cast as a thin film on silicon. The cationised fragments of the PEG side-chain (average molecular weight ~1000) are easily seen in the log (intensity) plot (a). A portion of this spectrum at high mass resolution (b) allows identification of the alternating fragments of generic composition $CH_3O(CH_2CH_2O)_nCH{=}CH_2$ (the methyl terminated PEG fragment) cationised by Na^+ and K^+ respectively (Briggs & Davies, 1997).

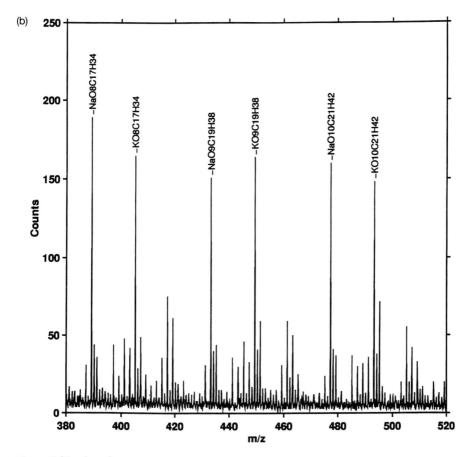

Figure 5.21   (*cont.*)

Chapter 6

Polymer surface analysis case studies

6.1   Introduction

Chapters 3 and 5 describe the information levels of XPS and SSIMS, respectively. The two techniques are highly complementary, having certain shared attributes but differing markedly in other aspects. XPS detects all elements except hydrogen, is inherently quantitative but has limited sensitivity (typically 0.2 atomic % detection limit for low photoelectron cross-section elements) whereas SSIMS detects all elements with isotope specificity (but some, notably nitrogen, only in combined ions), is inherently difficult to quantify but can have very high sensitivity (ppm of one monolayer). Structural information in XPS is dominated by functionality and limited by the poor dynamic range of chemical shifts whereas SSIMS has a high level of molecular specificity and can discriminate molecular from macromolecular entities. XPS typically has a sampling depth of $\sim$5 nm but in certain circumstances this can be varied and reduced to $\sim$1 nm. The SSIMS sampling depth is essentially fixed and corresponds to the outermost monolayer. Both techniques are capable of analysing small areas of the order of 50 μm width/diameter and of imaging with reasonable spatial resolution (XPS $\sim$10 μm, SSIMS $\sim$1 μm under typical conditions). From this it is clear, even in principle, that combining information from the two techniques will be beneficial.

In the early days of SSIMS, when the instrumentation was based on compact quadrupole mass spectrometers, combination XPS/SSIMS systems were popular. Both techniques could be used to examine a sample sequentially in a cost-effective way (only one vacuum system is needed and there are no uncertainties about comparing information from discrete samples). As the instrumentation

156

has evolved in complexity (especially with ToF SIMS) stand-alone systems have become the norm. However, an instrument has been developed which combines XPS and ToF SIMS through clever use of the lens–analyser–lens system either as an imaging electron spectrometer (XPS) or as an energy-compensating ToF mass spectrometer (SSIMS) (Coxon & McIntosh, 1996). More compact, moderate mass resolution, ToF analysers will doubtless appear in the future and reestablish the cost-effective XPS/SSIMS combination.

The aim of this chapter is to illustrate, through a series of selected examples, how XPS and SSIMS are used to investigate polymer surfaces and to understand surface behaviour in a variety of material contexts. Most of the examples have been chosen specifically to emphasise the benefits of combining XPS and SSIMS analysis.

6.2   Surface characterisation of EMA copolymers

EMA copolymers have the composition:

$$[(CH_2-CH_2)_x-(CH_2-CH)_y]_n$$
$$\begin{array}{c} | \\ C=O \\ | \\ OCH_3 \end{array}$$

They are semicrystalline, with ethylene enrichment of the crystalline phase and methyl acrylate enrichment of the amorphous phase, and commonly used in films and adhesive applications. The distribution, composition and orientation of the crystalline and amorphous phases at the surface are expected to have a major influence on the surface properties which are critically important in these applications (e.g. cling, sealing, wetability, printability and adhesion). A detailed study of a range of EMA copolymers with methyl acrylate content varying from 6.8 to 33 wt%, using XPS and ToF SIMS, has given the first detailed insight into the complex architecture of the surface region of EMA films (Galuska, 1996).

To probe a range of depths below the surface, variable angle XPS (take-off angles of 20°, 45°, 75° and 90°) was employed and to compare surface and bulk spectra the sample was melted *in situ* by heating to 180°C. Fig. 6.1 shows typical C1s and O1s spectra fitted with the appropriate components for the acrylate functionality, having the appropriate stiochiometry. Additionally some ether (C–O) and carbonyl (C=O) functionality is present due to surface oxidation (primarily of the ethylenic component). Fits to both peaks give a self-consistent result such that for the example shown (23.4 wt% methyl acrylate) the total of 7 at% oxygen at the surface is made up of 6 at% due to methyl acrylate and 1 at% due to adventitious oxidation. Since the latter is primarily as ether the easiest way

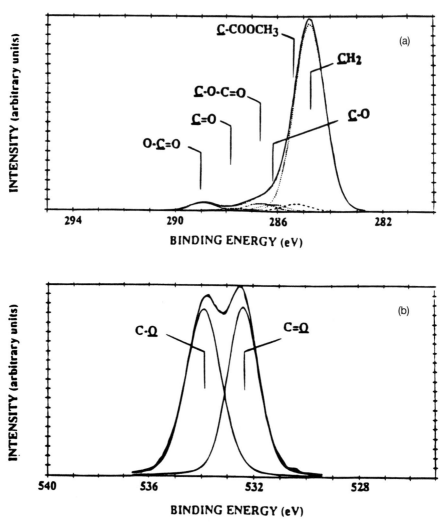

Figure 6.1 (a) C 1s and (b) O 1s spectra from the surface of a thermally pressed film of an EMA copolymer containing 23.4 wt% methyl acrylate (45° take-off angle) (Galuska, 1996).

to measure the surface MA oxygen concentration reliably is to double the O—C=O concentration (from the C 1s spectrum).

Fig. 6.2 thus plots the methyl acrylate oxgyen concentration as a function of bulk methyl acrylate concentration as determined from different depths below the surface. The data can be interpreted as follows. For EMAs with a bulk methyl acrylate content below 20 wt% the methyl acrylate concentration maximises at slightly above the bulk value in the top 2 nm and decreases to well below the bulk value with increasing analysis depth. Between 23–25 wt% bulk methyl acrylate the methyl acrylate concentration

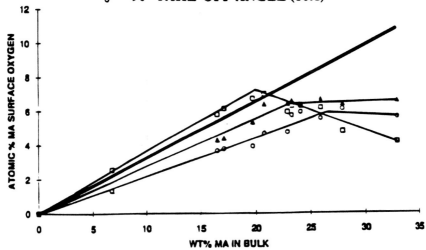

BULK MA OXYGEN CALIBRATION
□ **20° TAKE-OFF ANGLE (20Å)**
▲ **45° TAKE-OFF ANGLE (40Å)**
○ **90° TAKE-OFF ANGLE (60Å)**

Figure 6.2 Surface methyl acrylate oxygen concentration vs bulk methyl acrylate concentration for thermally pressed EMA films as determined from XPS measurements at three different take-off angles. The bold line would be obtained for a surface composition identical with the bulk (Galuska, 1996).

is similar for all analysis depths and significantly below the bulk value. For EMAs of above 25 wt% bulk methyl acrylate the methyl acrylate oxygen is depleted from the surface (2 nm), concentrates at a depth of ~4 nm and decreases below 6 nm. The periodicity of this layered structure is on a size scale inconsistent with segregation based on crystallinity. More likely is segregation of the amorphous phase into methyl-acrylate-rich and ethylene-rich layers due to clustering of methyl acrylate. Heating experiments show that the surface layering is completely removed by heating to 180°C and equilibrating for 30 min (XPS gives the bulk methyl acrylate concentration regardless of sampling depth).

Whilst XPS can quantify surface methyl acrylate concentrations, these are not uniquely characteristic of the bulk. This problem could be overcome if another technique determined whether the bulk concentration was above or below 20 wt% so that the 2 nm calibration line in Fig. 6.2 could be utilised to quantify bulk methyl acrylate concentration. Furthermore, it is worth emphasising that XPS is unable to identify an EMA copolymer positively from the core level spectra (although this might be possible from the valence band spectrum).

The SSIMS spectra of EMA copolymers are, at low mass resolution, lacking in peaks which identify methyl acrylate. As expected, the low methyl acrylate

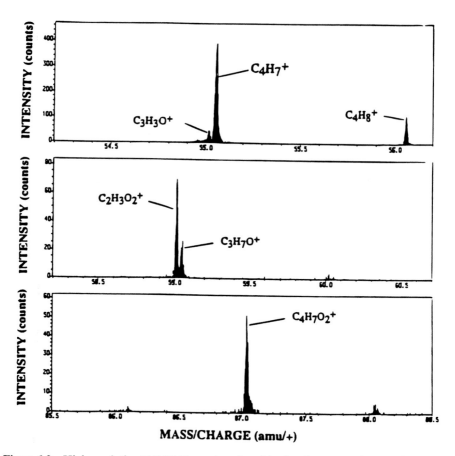

Figure 6.3 High resolution ToF SIMS spectra of positive ion fragments characteristic of the methy acrylate funcationality from an EMA film containing 17 wt% methyl acrylate (Galuska, 1996).

content copolymers give spectra similar to poly(ethylene) although the intensities of CH_3^+, $C_5H_9^+$, $C_6H_{11}^+$ are enhanced and these peaks become dominant as the methyl acrylate content increases. Whilst oxygen is detected the negative ion spectrum is uninformative. At high mass resolution the positive ToF SIMS spectrum reveals numerous low intensity oxygen-containing fragments characteristic of the methyl acrylate. Some of these are shown in Fig. 6.3. From a detailed study of the spectral changes across the copolymer series and an understanding of ion formation mechanisms it was concluded that ions like $C_7H_{13}^+$ represent the polymer backbone with attached methyl acrylate side-chains whilst ions like $C_2H_3O_2^+$ represent the side-chain itself. Fig. 6.4 shows relative intensity plots (normalising to the non-specific $C_2H_3^+$ intensity) for these ions as a function of bulk methyl acrylate concentration.

Fig. 6.4(a) shows the same behaviour as the 2 nm (20 Å) correlation in Fig. 6.2 but in an exaggerated form, emphasising the criticality of the copolymer

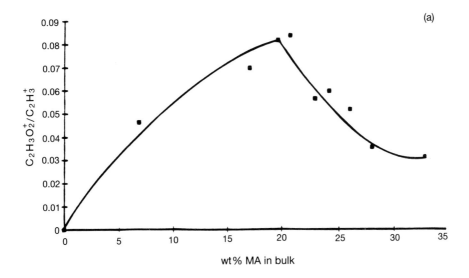

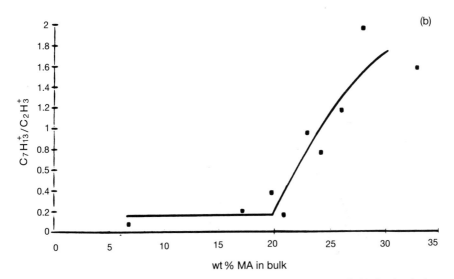

Figure 6.4   SSIMS intensity plots for the ions $C_2H_3O_2^+$ (a) and $C_7H_{13}^+$ (b) (both relative to $C_2H_3^+$) as a function of bulk methyl acrylate content for the series of EMA copolymers (Galuska, 1996).

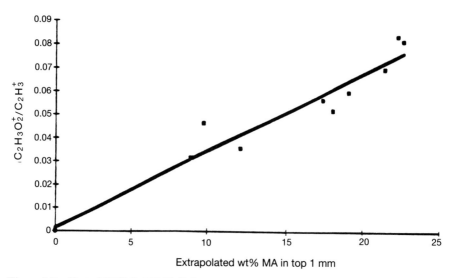

Figure 6.5   Plot of $C_2H_3O_2^+/C_2H_3^+$) from ToF SIMS vs concentration of surface methyl acrylate estimated by XPS (following extrapolation of angle-resolved data to a sampling depth of 1 nm – the presumed SSIMS sampling depth) (Galuska, 1996).

composition above or below 20 wt% and suggesting a correlation between the XPS and SSIMS measures of surface methyl acrylate concentration. If the XPS data are extrapolated to 1 nm sampling depth so as to coincide with the SSIMS sampling depth the correlation is excellent (Fig. 6.5). Fig. 6.4(b) shows that above 20 wt% methyl acrylate in the bulk the backbone fragments suddenly increase in intensity. This strongly suggests a rearrangement of polymer chains at the surface so that the ester side-chains are preferentially orientated inwards.

ToF SIMS can therefore supply the knowledge of whether the copolymer is in the less than or greater than 20 wt% bulk methyl acrylate category. Hence XPS can then be used to determine quantitatively both the bulk methyl acrylate concentration and the variation of composition with the upper 6 nm or so.

6.3   Surface characterisation of biomedical poly(urethanes)

As was noted in Chapter 1, the usage of polymers in biomedical applications is already extensive and still growing rapidly. In most of these applications the nature of the device surface is very important since it determines the biological response. One important class of polymers in this field is the segmented poly(etherurethanes), the generic synthesis of which is given in Fig. 6.6. The poly(ether) unit, typically having a molecular weight in the range 400–2000,

Polyetherurethane Synthesis

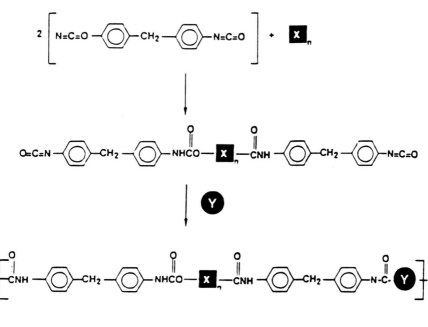

X	Abbreviation	
-(-CH₂CH₂-O-)ₙ	PEG	
$\overset{\displaystyle CH_3}{\underset{\displaystyle}{}}$ -(-CH₂CH-O-)ₙ	PPG	
	DPG TPG	polyether soft segment
-(-CH₂CH₂CH₂CH₂O-)ₙ	PTMG	
	BD unit	

Y	Abbreviation	
NH₂CH₂CH₂NH2	ED	chain extender
OHCH₂CH₂CH₂CH₂OH	BD	

Figure 6.6 Poly(etherurethane) synthesis and structures of typical soft segment polyethers and chain extenders (Hearn, Ratner & Briggs, 1988).

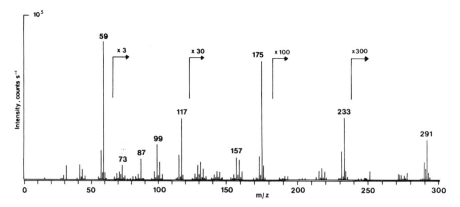

Figure 6.7   Positive ion quadrupole SSIMS spectrum of PPG, molecular weight 425 (Hearn *et al.*, 1988).

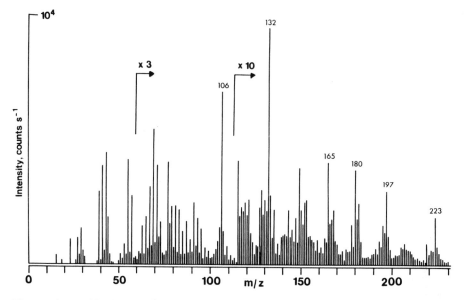

Figure 6.8   Positive ion quadrupole SSIMS spectrum of a butanediol hard segment model poly(urethane) (Hearn *et al.*, 1988).

forms the 'soft segment' (flexible chain) of the poly(etherurethane) whilst the di-isocyanate, in this case MDI, plus chain extender (a diamine or diol) forms the rigid 'hard segment'. Variation of the components and their relative concentrations provides the means to generate materials covering a wide property spectrum. These polymers have a tendency (dependent on composition and processing history) to form discrete phases, with domain sizes of 10–20 nm, in the bulk. Thus an understanding of biocompatability requires a description of surface structure including lateral and vertical heterogeneity.

Although the pure poly(ethers) can be disginguished by XPS (e.g. Beamson &

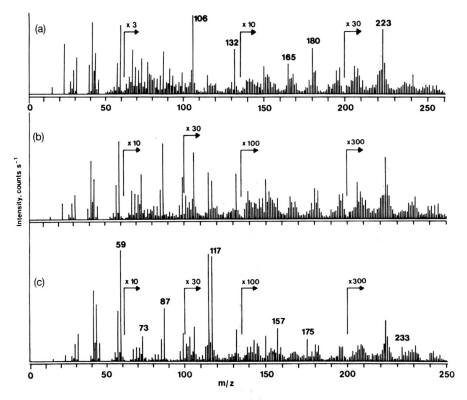

Figure 6.9   Positive ion quadrupole SSIMS spectra of PPG/MDI/ED PEUs prepared with (a) PPG 425, (b) PPG 775, (c) PPG 2000 (Hearn *et al.*, 1988).

Briggs, 1992a) their identification in the poly(urethanes) is very difficult. However, their SIMS 'signatures' (the set of unique peaks in the positive ion spectra) are carried over into the poly(urethane) spectra. By way of example, Figs. 6.7–6.9 show quadrupole SSIMS spectra of poly(propylene glycol) (PPG), a hard segment model polymer in which butanediol replaces the poly(ether) and a set of poly(etherurethanes) based on PPG, MDI and ethylene diamine (the chain extender) in which the PPG molecular weight varies. Fig. 6.9 shows that, as the molecular weight of the PPG soft segment increases, the SSIMS spectrum is dominated by the polyether peaks. Partly this is due to the increasing fraction of PPG in the polymer, as shown by curve (a) in Fig. 6.10, where peaks at *m/z* 59 and 106 are uniquely associated with the soft and hard segments, respectively. However, curve (b) shows that on a molar ratio basis increasing the soft segment concentration dramatically increases the soft segment peak intensities, indicating surface segregation of PPG units (Hearn *et al.*, 1988). It is interesting to note, in passing, that the relative intensities of the PPG fragment ions in the pure poly(ether) and in the poly(urethane) are not the same (either an end-group effect or a result of conformational restrictions) and this allows the possibility

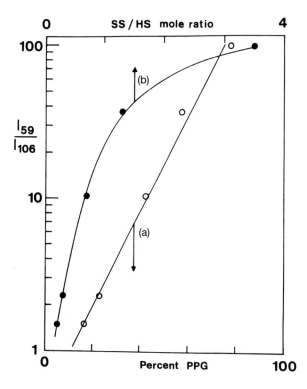

Figure 6.10 Log (I_{59}/I_{106}) peak intensity ratios plotted as a function of bulk concentration expressed in terms of (a) % PPG content and (b) soft segment (SS) to hard segment (HS) ratio for PPG/MDI containing PEUs (Hearn *et al.*, 1988).

that the segregating PPG units are 'free' PPG homopolymer to be disproved. Studies of polymers either solvent extracted or doped with pure PPG provide further confirmation. On the other hand SSIMS cannot (at least on the basis of low mass resolution spectra) identify the chain extender unit in the poly(urethane), but XPS can. Thus, in Fig. 6.11 the butanediol containing polymer only contains carbamate (NH—CO—O) linkages whilst the ethylene diamine containing polymer contains carbamate and urea (NH—CO—NH) linkages. These two groups give slightly different C1s binding energies so that the high binding energy peak is significantly broader in (b) and capable of being resolved into two components (not shown).

Surface segregation in these systems has been studied by ARXPS. This has to rely on quantitative nitrogen analysis for hard segment concentration determination. For the PPG–PEU with the highest soft segment/hard segment ratio discussed above the nitrogen content within the maximum XPS sampling depth is <2 at%. Decreasing the sampling depth reduces the nitrogen concentration, in confirmation of the SSIMS results, but also leads to a significant reduction in signal:noise with most XPS instruments (due to the low take-off angle). Since the nitrogen detection limit is ~0.3 at%, low take-off angle data can easily lead to the conclusion that the surface layer is pure soft segment. The greater sensitivity of SSIMS to hard segment shows this is never actually the case, which has

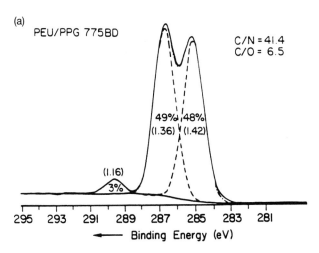

(a) PEU/PPG 775BD

C/N = 41.4
C/O = 6.5

49% 48%
(1.36) (1.42)

(1.16)
3%

295   293   291   289   287   285   283   281

◄── Binding Energy (eV)

Figure 6.11 C 1s XPS spectra of (a) PPG775/MDI/BD and (b) PPG775/MDI/ED PEUs. Peak fwhm values are indicated in parentheses. Note the significant difference in the width of the $>C{=}0$ component at $\sim$289.5eV (Hearn *et al.*, 1988).

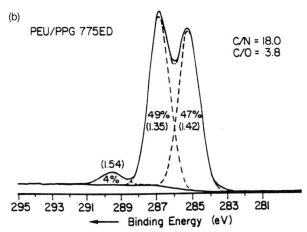

(b) PEU/PPG 775ED

C/N = 18.0
C/O = 3.8

49% 47%
(1.35) (1.42)

(1.54)
4%

295   293   291   289   287   285   283   281

◄── Binding Energy (eV)

important implications for constructing models of the polyurethane surface microstructure.

With this background we turn to the study of Biomer surfaces. Biomer is a commercial, medical-grade poly(urethane) of undisclosed composition. Variation in the properties of different batches of the material have been observed over time and compositional variations detected by pyrolysis mass spectra, although analysis by FT–IR spectroscopy showed the bulk composition to be the same.

The surfaces of two distinct batches of Biomer, referred to as BSP and BSUA have been studied by XPS and quadrupole SSIMS (Tyler *et al.*, 1992). The XPS spectra are shown in Figs. 6.12 and 6.13 and the SSIMS spectra in Figs. 6.14 and 6.15. Clearly the surfaces of the two materials are very different. Based on the background studies described above, the XPS and SSIMS data for BSP are consistent with a relatively pure surface of a segmented poly(etherurethane) based on poly(tetramethylene glycol) (PTMG), MDI and ethylene diamine. In the SSIMS

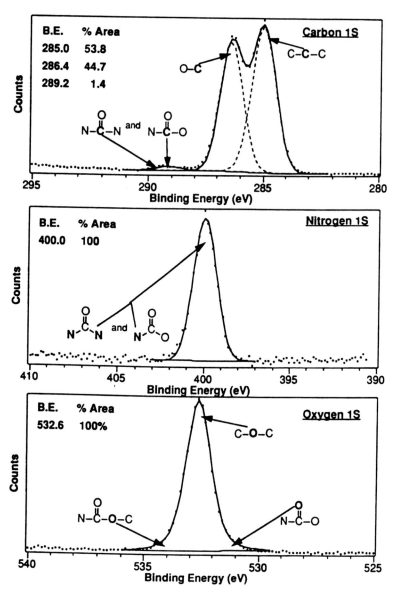

Figure 6.12 Core level XPS spectra from batch BSP. In the O 1s spectrum, contributions from amide oxygens at binding energies either side that of the ether oxygen are too small to be accurately fitted (Tyler *et al.*, 1992).

spectrum (Fig. 6.14) the peaks at *m/z* 55, 71 and 127 are characteristic of PTMG, whilst those at *m/z* 106 and 132 represent the (MDI-based) hard segment. Unexpected peaks appear at *m/z* 147, 161 and 177. These are characteristic fragments of the hindered phenol antioxidont Irganox 245 (a higher mass range ToF SIMS instrument would have detected the molecular ion at *m/z* 587 ([M+H]⁺), which illustrates the power of SSIMS to identify low concentration surface molecules.

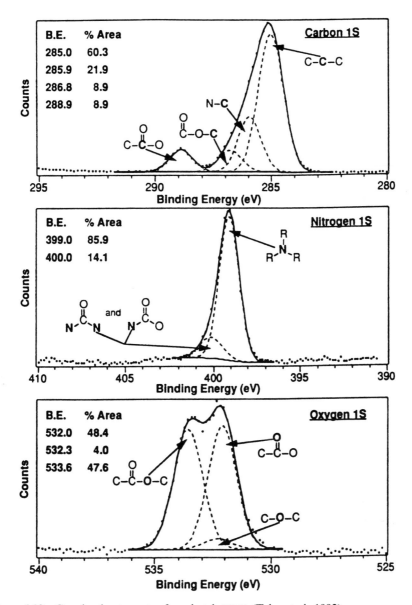

Figure 6.13 Core level XPS spectra from batch BSUA (Tyler *et al.*, 1992).

The SSIMS spectra from the BSUA batch are also most informative. None of the peaks seen in Fig. 6.14 is present and the negative ion spectrum, which is usually uninformative for poly(urethanes), is rich in peaks. Both spectra are consistent with the presence of a di-isopropyl amino ethyl methacrylate (DPA–EMA) containing material, based on a systematic interpretation of the major peaks and knowledge of the spectra from a dimethyl amino ethyl mathacrylate copolymer (Wilding *et al.*, 1990). With the SSIMS insight, the XPS data (Fig. 6.13) can be fully exploited since curve fitting is based on known functionality.

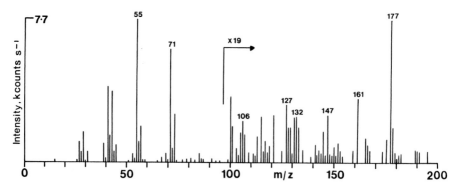

Figure 6.14   Positive ion quadrupole SSIMS spectrum from batch BSP (Tyler *et al.*, 1992).

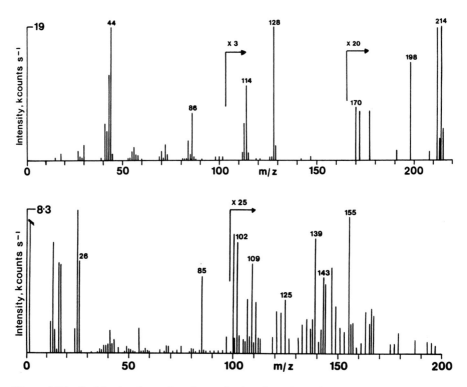

Figure 6.15   Positive ion (upper) and negative ion (lower) quadrupole SSIMS spectra from batch BSUA (Tyler *et al.*, 1992).

Exhaustive extraction experiments failed to convert the BSUA surface into a more BSP-like surface, hence the DPA–EMA component is firmly bound into the surface structure. Literature controversy relating to the biological interactions of Biomer may be partly explained by these major surface chemical differences and the necessity of a full surface characterisation of such polymers prior to biological testing is indicated.

6.4 Surface modification of polyalkenes by electrical discharge treatment

Electrical ('corona') discharge treatment of poly(ethylene) and poly(propylene) films has long been used to render these surfaces printable and suitable for lamination or coating. In this process the polymer film is passed over an earthed metal roller covered with a thick layer or an electrically insulating material. Separated from the film by ~2mm is an electrode bar to which a high voltage is applied (typically 15kV at 20kHz). Air in the film–electrode gap is ionised, the corona discharge thus formed is stable and this 'treats' the film surface. The most easily measured change in surface properties is that of surface energy (from contact angle) which is increased significantly from the initial value of ~31mNm^{-2}. The level of treatment depends on, amongst other things, the power dissipated in the discharge and the film speed. Laboratory experiments to simulate the process are usually carried out on small-scale static rigs with parallel metal plates replacing the roller and electrode bar and at much lower frequency (50Hz).

Although discharge treatment had been in use for 20 years prior to the advent of the surface analysis techniques discussed here, little was known about the nature of the surface modification process or why this produced the desired improvements in adhesion characteristics. The application of XPS and, more recently, SSIMS to this problem gives an interesting historical perspective on the relationship between analytical capability and the level of understanding achieved.

When XPS was first invoked the aspect of discharge treatment which had been most studied was that of the enhancement of low density polyethylene (LDPE) autoadhesion. LDPE film autoadheres when two surfaces are contacted under pressure at temperatures above ~90°C. However, after fairly low levels of discharge treatment, treated surfaces will autoadhere at significantly lower temperatures (~70°C). Two theories had been proposed to explain this effect. The first, due to Canadian workers (Stradal & Goring, 1975), suggested that electret formation was involved – the increase in adhesion being electrostatic in nature. The evidence for this was that discharge treatment in both 'active' (air, oxygen) and 'inert' (nitrogen, argon, helium) gases gave the effect, its magnitude being related to the power dissipated in the discharge irrespective of the gas, and that the maximum effect was achieved in oxidising gases before reflection infrared spectroscopy could detect any polymer oxidation. The second, diametrically opposed theory (Owens, 1975) suggested that hydrogen bonding between polar groups formed at the surface was responsible, but only on the basis of experiments involving air discharge treatment under full-scale plant conditions.

The first XPS experiments set out to resolve this controversy. Some of the critical results are shown in Fig. 6.16 (Blythe et al., 1978). For all the gases used oxygen was detected in the surface and the amount correlated with the degree of

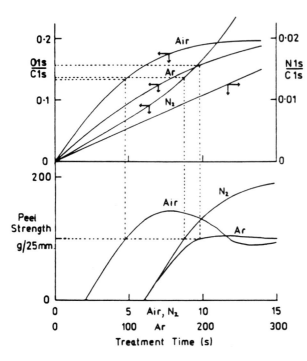

Figure 6.16 Comparison of autoadhesion (peel strength) and surface composition (from XPS) for LDPE discharge treated in air, nitrogen and argon as a function of discharge treatment time (in a static cell operated at 50 Hz with peak voltages for the gases of 13.7, 12.7 and 2.2kV, respectively). Heat seals were made at 75°C and 15lbin^{-2} with a 2s contact time. The O 1s:C 1s intensity ratio is a qualitative measure of surface oxidation level. The N 1s:C 1s ratio is for surfaces treated in nitrogen only. Note the similar oxidation levels for samples giving peel strengths of 100g/25mm, the highest value achieved for Ar treatment (indicated by the broken lines) (Blythe *et al.*, 1978).

autoadhesion (peel strength) irrespective of the conditions required to generate this. In all other respects the results confirmed the Canadian experiments. However, discharge treatment in hydrogen did not lead to enhanced autoadhesion and XPS did not detect surface oxidation. Thus Owens' theory was given a firmer foundation, but further support would require a much more detailed knowledge of the surface functionality.

Representative core level spectra from LDPE treated in air are given in Fig. 6.17. Two aspects need to be emphasised. First, commercial discharge treatment conditions leading to acceptable surface behaviour give spectra similar to (b) whereas (c) represents severe overtreatment. Second, even when the high binding energy shoulder is sufficiently developed to allow curve fitting the insight into surface functionality is very limited. Thus C—O (~286.5eV) could represent alcohol, ether, ester or hydroperoxide, C=O (~288.0eV) could represent aldehyde, ketone or ozonide and O—C=O (~289.5eV) could represent carboxylic acid or ester. The O 1s peak is even less informative, as discussed in Chapter 3. A series of derivatisation reactions (see Section 3.8) was devised to probe specified functional groups (Briggs & Kendall, 1982) and these results appear in Table 6.1.

The value for the population of CH$_2$C=O groups assumes that both α-H atoms are replaced on bromination. This group can tautomerise to give one

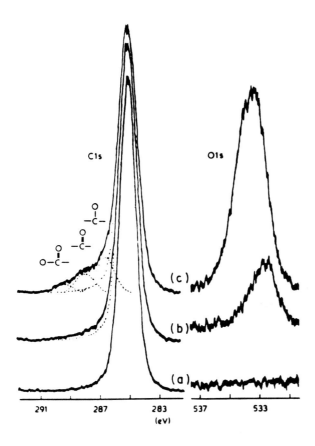

Figure 6.17 C 1s and O 1s spectra for LDPE: (a) untreated; (b) and (c) discharge treated in air (13.7 kV peak voltage, 50 Hz) for 8 s and 30 s respectively. The emerging high binding energy shoulder can be fitted with three components representing, in general, carbon atoms having one, two and three bonds to oxygen (Blythe *et al.*, 1978).

Table 6.1 *Quantification of functional groups in a corona discharge treated, LDPE surface, using derivatisation reactions/XPS (Briggs & Kendall, 1982)*

Reaction	XPS ratio (core level/C 1s)	Atomic ratio (element/carbon)	Number of functional groups per original surface $-CH_2-$
PFPH	(F 1s) 0.205	5.5×10^{-2}	$>C=O$, 1.1×10^{-2}
Br_2–H_2O	(Br 3d) 3.6×10^{-2}	10.6×10^{-3}	$CH_2C=O$, 5.3×10^{-3}
CAC[a]	(Cl2p) 1.3×10^{-2}	6.0×10^{-3}	$C-OH$, 6.0×10^{-3}
TAA[b]	(Ti2$p_{3/2}$) 6.2×10^{-2}	1.5×10^{-2}	$C-OH$, 1.5×10^{-2}
NaOH	(Na 1s) 8.8×10^{-2}	1.1×10^{-2}	$-COOH$, 1.1×10^{-2}
SO_2	(S2p) 7.6×10^{-3}	4.7×10^{-3}	$C-OOH$, 4.7×10^{-3}
None	(O 1s) 0.209	8.7×10^{-2}	

[a]Reacts apparently selectively with enolic $-OH$.

[b]Reacts apparently selectively with alcoholic $-OH$.

 PFPH=pentafluorophenylhydrazine, CAC=chloroacetylchloride.

 TAA=di-isopropoxytitanium bisacetylacetonate.

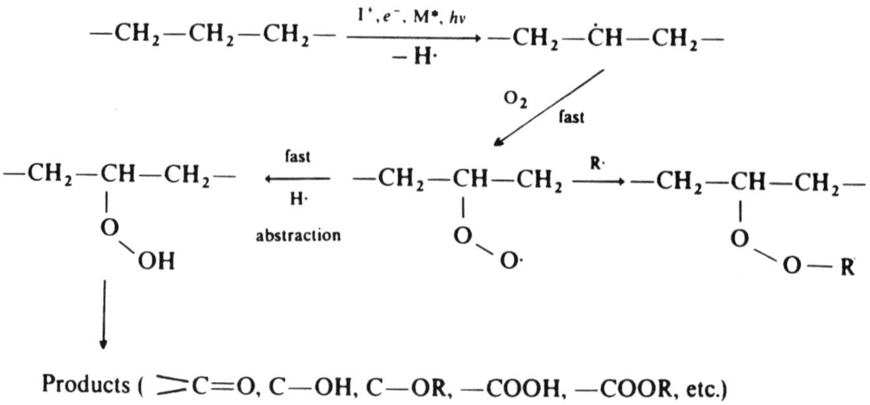

$$-CH_2-CH_2-CH_2- \xrightarrow[-H\cdot]{I^+, e^-, M^\bullet, hv} -CH_2-\overset{\displaystyle\cdot}{C}H-CH_2-$$

$$O_2 \searrow \text{fast}$$

$$-CH_2-CH-CH_2- \xleftarrow[H\cdot]{\text{fast}} -CH_2-CH-CH_2 \xrightarrow[]{R\cdot} -CH_2-CH-CH_2-$$

abstraction

$$\underset{\displaystyle OH}{O}\qquad\qquad \underset{\displaystyle O\cdot}{O}\qquad\qquad \underset{\displaystyle O-R}{O}$$

Products ($>C=O$, C—OH, C—OR, —COOH, —COOR, etc.)

Figure 6.18 Likely mechanism of oxidation of LDPE by discharge treatment. Hydroperoxide decomposition is frequently accompanied by chain scission leading to functionalised products of reduced molecular weight.

enolic —OH, which can be derivatised using chloroacetyl chloride. The two procedures lead to a comparable quantification. Overall the results are in broad agreement with the curve fitting approach, given the groups which cannot easily be assessed (e.g. carboxylic ester), and are entirely consistent with the free radical oxidation mechanism shown in Fig. 6.18. The key intermediate is the hydroperoxide group which can have a long lifetime in poly(ethylene), and is detectable by SO_2 derivatisation.

 Printing inks do not adhere to untreated LDPE (or poly(propylene)). The relationship between LDPE surface chemistry (from XPS analysis), printability (ink adhesion from a conventional 'sellotape' peel test) and the degree of discharge treatment is presented in Fig. 6.19. The printability increases rapidly following the attainment of a threshold level of oxidation – most likely that required to increase the surface energy sufficiently to give good wetting by the ink solvents. Derivatisation with pentafluorophenylhydrazine (PFPH) shows that the carbonyl concentration is a roughly constant fraction of the total oxygen functionality over the whole treatment range. This reaction, and others that eliminate the possibility of enolic —OH formation, reduce ink adhesion to zero; hence, wettability is a necessary but not sufficient condition for printability. (Autoadhesion is also reversed so these results confirm the original Owens theory which postulated adhesive interactions based on strong hydrogen bonding through the enolic —OH group.) The oxidation mechanism of discharge treatment leads to chain scission and, hence, low molecular weight oxidised material at the surface. The gap between curves (b) and (c) in Fig. 6.19, must represent, to some extent, the relative concentration of low molecular weight oxidised material. This material, being cohesively weak, is a potential 'weak boundary layer' which could reduced adhesion (e.g. to a high solids content UV-curable ink) and a knowledge of its presence is very useful. In the case of the

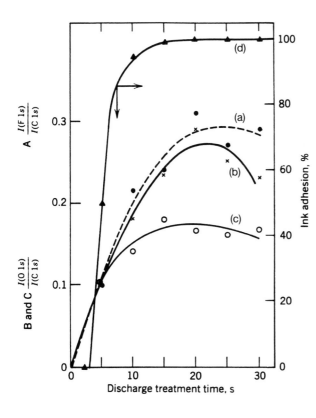

Figure 6.19 XPS core level relative intensity data and printability of LDPE films as a function of discharge treatment level (seconds exposure under constant discharge conditions in a static treatment cell). Curves (b) and (c) refer to the O 1s:C 1s intensity ratio after treatment and after further derivatisation of carbonyl groups with PFPH in ethanol, respectively. Curve (a) is the F 1s:C 1s intensity ratio for the same surface as for (c). The printability curve (d) is from the same surface as for (b). Although the derivatisation reaction itself results in the loss of oxygen from the surface, most of the reduction in the O 1s:C 1s ratio is through ethanol dissolution of low molecular weight oxidised material (after Ashley *et al.*, 1985).

liquid ink used in the above experiments the low molecular weight oxidised material can be absorbed into the ink before it cures.

In discharge treatment of poly(propylene) chain scission is more likely than in the case of LDPE (Briggs *et al.*, 1983) and low molecular weight oxidised material production even more important. The ageing behaviour of discharge treated surfaces (changes in surface energy and adhesion behaviour) as a function of treatment level has led to speculation about the role of the low molecular weight oxidised material. This has been studied by AFM, XPS and SSIMS (Boyd *et al.*, 1997). For poly(propylene) treated to higher levels of oxidation than Fig. 6.17(c), AFM revealed regular globular structures of ~0.5–1 μm diameter which were largely removed by solvents. High mass resolution ToF SIMS was used to study the poly(propylene) surface after discharge treatment and the material removed by exposure to chloroform. At low mass resolution the positive SSIMS spectrum is not radically different from that of the untreated surface. However, closer inspection of the peaks at any nominal mass with high mass resolution reveals oxygenated fragments in addition to the original hydrocarbon ($C_xH_y^+$) fragments. Some examples are shown in Fig. 6.20. Interestingly one of the few new peaks to appear after discharge treatment is that at *m/z* 113,

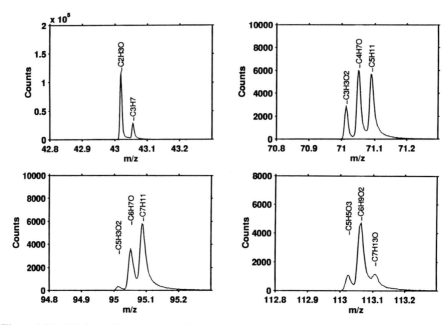

Figure 6.20   High resolution positive ion ToF SIMS spectra from discharge treated poly(propylene). The chosen peaks are representative of the appearance of mass resolved, oxygen containing fragments having the same nominal mass as the purely hydrocarbon fragments present in the untreated polymer spectra (Boyd *et al.*, 1997).

consistent with all the component fragments being oxygenated. A much more striking change is seen in the negative ion spectrum. An extensive series of fragment clusters extending almost to m/z 500 is generated by discharge treatment. Fig. 6.21 shows the exponentially decreasing intensity of the clusters of peaks. All the peaks are due to ions with the general structure $C_xH_yO_z^+$ and at each nominal mass there are several components. Fig. 6.22 illustrates this for just one of the clusters of peaks in the region of m/z 150–160. The fragments represented have the generic formulae $C_6O_5H_n^-$, $C_7O_4H_n^-$, $C_8O_3H_n^-$ and possibly $C_5O_6H_n^-$ and $C_9O_2H_n^-$ (weaker components). The results confirm the very high O:C atomic ratio measured by XPS. Also unambiguous identification of CO_3^- and HCO_3^- in the negative ion spectrum helps to confirm the assignment of the highest binding energy component of the C 1s spectrum as the carbonate function.

Material extracted from the surface by brief solvent exposure was deposited on etched silver. By trial and error adjustment of the solution concentration the appropriate coverage for cationisation was achieved, giving the spectrum shown partially in Fig. 6.23. There are three sets of peaks: those due to the silver substrate, those due to the antioxident Irganox 1076 (m/z 637/639 from $[M+Ag]^+$) – detected despite rigorous attempts to extract it from the poly(propylene) film prior to discharge treatment – and an envelope of peaks between ~m/z 500–1400. Each component of the envelope is composed of a pattern of peaks (with very similar intensity ratios) with a repeat interval of mass equivalent to

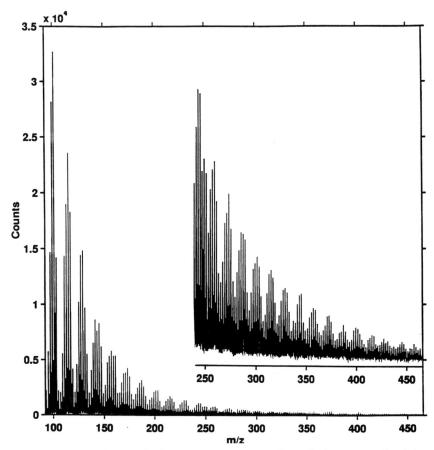

Figure 6.21 Part of the negative ion ToF SIMS spectrum from discharge treated poly(propylene) (Boyd *et al.*, 1997).

CH_2. Within each pattern the two most intense peaks are separated by a mass equal to $^{109}Ag-^{107}Ag$ and these can be assigned the generic formulae $[(CH_2)_n OAg]^+$ on the basis of accurate mass measurement. The envelope may therefore be due to low molecular weight oxidised material resulting from oxidative chain scission of the poly(propylene) backbone yielding a single functional chain end. The highest clusters in Fig. 6.23 correspond to chains involving 100 carbon atoms. It seems likely that there exists a wide range of entities with varying composition and molecular weight in the surface region affected by discharge treatment whose different potentials for interaction, diffusion and segregation will lead to a highly dynamic situation.

6.5 Surface morphology of PVC/PMMA blends

Through the blending of polymers it is possible to obtain material properties not available to the single constituents. Polymer blends are, therefore, technologically

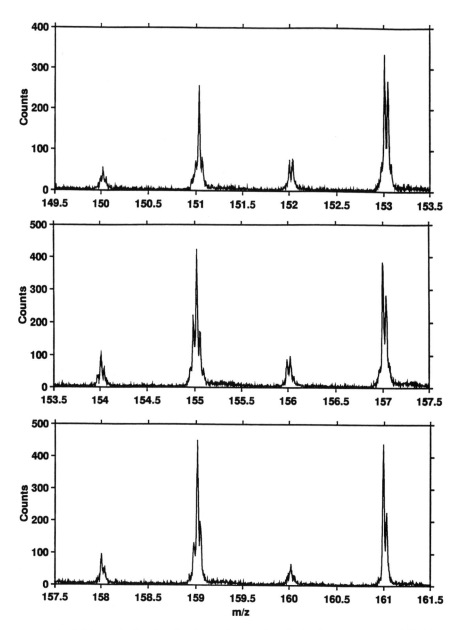

Figure 6.22   High resolution details of one cluster of peaks in the mass range ~150–160 shown in Fig. 6.21. All the peaks appearing in the negative ion spectrum of discharge treated poly(propylene) are due to the oxygen containing fragments (Boyd *et al.*, 1992).

extremely important and the thermodynamic and bulk physico-mechanical properties have been extensively studied (Walsh, Higgins & Maconnachie, 1985). Most binary polymer pairs are immiscible and produce bulk phase-separated materials on mixing; however, the surface morphology has received relatively little attention. For such a blend three extreme alternatives are possible:

(1) bulk morphology extended to the surface;
(2) molecular mixing of the two components in the surface region;
(3) preferential adsorption (segregation) of one component at the surface.

Distinguishing between these alternative surface morphologies may not be easy: it implies the simultaneous identification of chemical species at both high lateral and depth resolution within the surface region.

The study of PVC/PMMA blends is a good example. Several reports, using, amongst other techniques, ARXPS, had come to all possible conclusions: the surface had the same composition as the bulk; PVC surface segregated; PMMA surface segregated. Such controversy is not helped by the fact that different sample preparation conditions may well lead to different results. Important variables are the choice of cosolvent, the casting (spin casting, solution evaporation) and residual solvent removal procedures, and annealing conditions. Thermodynamically, and at equilibrium, the component with the lowest surface energy is predicted to enrich the surface. In the case of PVC and PMMA, the latter has the slightly lower surface energy (difference $<1\,\mathrm{mN\,m^{-2}}$).

Kunze et al. (1996) used neutron reflectometry (sampling depth $\sim$300nm with a depth resolution of <1nm), ARXPS and SSIMS to study the PVC/PMMAd (deuterated PMMA) system in two ways. Firstly, bi-layer samples were assembled from spin cast films (30–50nm thick) of the individual polymers and annealed for various times at temperatures of 108°C and 113°C. Even below T_g, diffussion of PMMAd into and eventually through the PVC layer was observed. At equilibrium the results from all three techniques agreed with an exponential (decreasing) concentration profile for PMMA away from the surface. Secondly, annealing experiments on blends with 5wt% PMMAd spin cast from cyclopentanone were performed. Similar results were obtained. In this case it seems to have been assumed that the bulk was homogeneous.

This is not the case for blends solution cast from tetrahydrofuran (THF); complex domain formation is easily observed by low resolution optical microscopy. This bulk inhomogeneity is typical of polymer blends and it makes the interpretation of ARXPS data rather difficult since this is based on a laminar overlayer–substrate model (see Sections 2.2.2 and 3.9). In order to resolve the different conclusions arrived at from such measurements, Jackson & Short (1992) turned to ToF SIMS imaging. Using a 30keV Ga$^+$ primary beam capable of sub-micron resolution, chemical images were obtained based on Cl$^-$ and

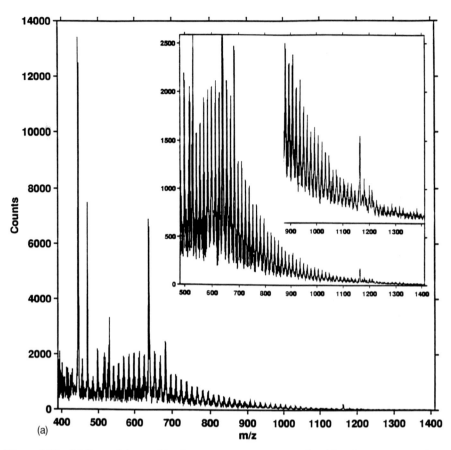

Figure 6.23 (a) Part of the positive ion ToF SIMS spectrum of a chloroform extract of discharge treated poly(propylene) deposited as a sub-monolayer on etched silver foil showing the cationised oligomer distribution between m/z 500–1400 (the peak at $\sim m/z$ 1160 has not been assigned). (b) Detail of representative repeating clusters of peaks each of which is seen as an unresolved component of the oligomer distribution in Fig. 6.23(a) (Boyd *et al.*, 1997).

$(O+OH)^-$. These clearly reproduced the domain morphology seen by scanning electron microscopy (SEM) and resolved PMMA domains as small as a few microns within a PVC continuous phase and vice versa. It was concluded that the bulk and surface morphology were the same. However, there is still ambiguity in these results. Only atomic or quasi-atomic secondary ions were studied. Whereas Cl^- is specific to PVC, there could be other contributors to O^-/OH^- than just PMMA (although unlikely). The highly characteristic molecular fragments in the negative ion spectrum of PMMA could not be used for imaging because their intensity was too low (almost certainly the result of inadequate charge neutralisation). Finally, the sampling depth of the secondary ions imaged could be significantly greater than the $\sim1\,nm$ associated with cluster ion emission in SSIMS.

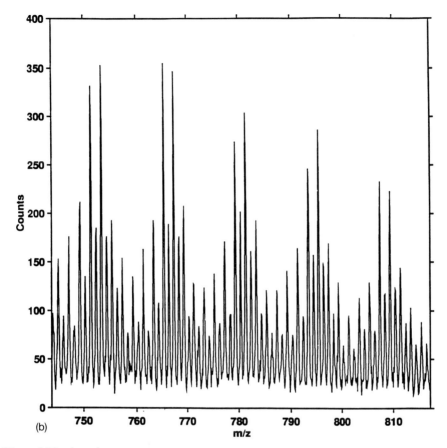

(b)

Figure 6.23   (*cont.*)

XPS imaging was carried out on identical samples prepared by solution evaporation (Briggs, 1993). Two colour overlay images based on Cl2p and O1s (peak-background in each case) gave exactly the same impression as the ToF SIMS images, namely an apparently clear reproduction of the bulk domain structure at the surface. However, inspection of the individual images (Fig. 6.24) shows this may be an illusion. Whilst the two images are complementary, the contrast (difference in intensity between regions of high and low intensity) is much greater in the Cl2p image. High intensity line-scans on an instrument capable of E–X imaging subsequently confirmed this impression. Assuming PMMA is the only source of O1s intensity, this suggests a thin PMMA overlayer (covering the domain structure extending almost to the surface). A retrospective imaging/shallow depth profiling ToF SIMS study of 50:50 PVC/PMMA blend has gone a long way towards resolving these inconsistencies (Briggs *et al.*, 1996). Fig. 6.25 shows images taken from the same $400 \times 400\,\mu m$ area under static conditions (total dose $< 10^{12}$ ions cm^{-2}) before and after a sputtering dose

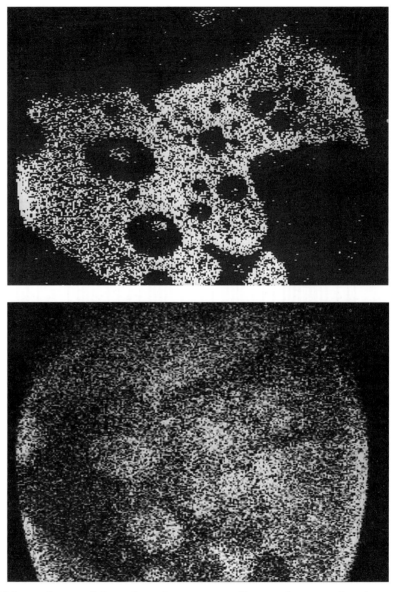

Figure 6.24 XPS images of the surface of a PVC/PMMA film, cast from THF, based on the Cl 2p (peak-background) intensity (upper) and the O 1s (peak-background) intensity (lower). The field of view is $700\,\mu$m (Briggs, 1993).

capable of removing several monolayers of material. In contrast with the earlier ToF SIMS study it was possible to image the characteristic negative ions of PMMA (m/z 31, 85, 87, 141 and 185 – see Section 5.2.2). The total negative ion image of the original surface clearly shows evidence of the domain structure. The PMMA-specific image is featureless but the Cl$^-$ image shows some evidence

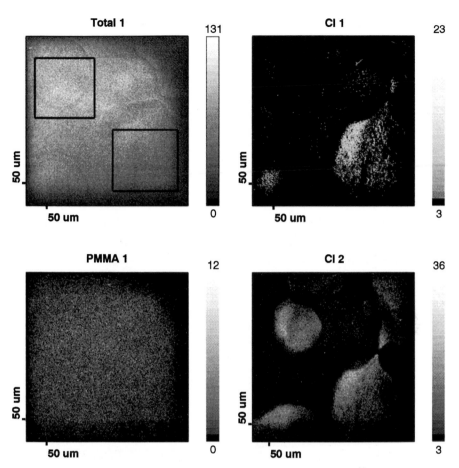

Figure 6.25 ToF SIMS images of the surface of a PVC/PMMA film cast from THF. Total 1 is the total negative ion intensity image in which phase boundaries can be discerned as a result of topographical contrast. Cl 1 and PMMA 1 are constructed retrospectively from the original datafile containing the full spectrum at image pixel using intensities of peaks representing chlorine and PMMA respectively. Cl 2 is a static image of the same area following a sputtering dose calculated to remove several monolayers. There is slight shifting and contraction, relative to the initial image, as a result of residual charging following the high ion dose. The intensity scale bars depict the number of ion counts per pixel in each image (Briggs *et al.*, 1996).

of the domain structure. After sputtering, this structure is developed with one new PVC domain in particular appearing (upper left). Retrospective micro-analysis was performed on the original total ion image regions corresponding to a PVC domain evident in the original Cl⁻ image and to the domain evident after sputtering. These are the boxed areas in image 'Total 1'. The spectra obtained by summing all pixels in each box gave only the characteristic PMMA spectrum plus the Cl⁻ isotopes, the intensity of the latter being relatively greater in the lower left box as expected from the PMMA 1 and Cl 1 images. These results show conclusively that a thin layer of PMMA uniformly covers the surface of the blend and that no other species are present (therefore in XPS there is no other contributor to the O 1s image but PMMA). The molecular fragment image from ToF SIMS is, of course, much more surface specific that the O 1s XPS image.

That chlorine can be detected in the original surface is probably due to the greater emission depth of the elemental (Cl^-) ion. An interesting piece of evidence which supports this interpretation is as follows: in the negative ion spectrum of PVC there is a significant Cl_2^- pattern (under the spectrometer conditions used for the blend experiments the $^{35}Cl^-/^{70}Cl_2^-$ intensity ratio is ~40) whereas even in the spectrum from the lower right box, which must represent a region where a PVC domain is closest to the surface, the Cl_2^- pattern cannot be detected. Because of its larger size Cl_2^- would be expected to have a lower emission depth than Cl^- but it is also a fragment formed by an addition process, most likely in the selvedge region and closer to the surface than the average collision cascade depth within which Cl^- is generated. In either case Cl_2^- would be suppressed by the presence of an overlayer.

Also relevant to these studies is the discovery that SSIMS can be very sensitive to retained solvent in cast polymer films. Leeson *et al.* (1997) investigated PMMA films cast from various solvents and the temperatures required to remove residual solvent. In the case of THF, even annealing well above the polymer T_g (105°C for PMMA) failed to remove all the solvent. This was evident from the SSIMS spectral intensities, but high mass resolution was required to explain why (Fig. 6.26). XPS was insensitive to the retained solvent. If there is a strong Lewis acid–base interaction between THF and PMMA, then this result could explain why miscible PMMA–PVC blends cannot be prepared from this solvent.

6.6   Electroactive polymers

In the very early days of the application of XPS to polymer characterisation and, indeed, before the surface sensitivity of the technique was fully appreciated, a major advantage was perceived to be its ability to study materials in the as-received state – thus obviating problems of intractability associated with insoluble, highly cross-linked (etc.) systems. A group of polymers which tend to fall into this category have become increasingly important during the last 20 years: electroactive ('conducting') polymers. These can be prepared so as to give electron conductivity and some electrical characteristics reminiscent of metals, although the precise nature of the charge carriers is still not fully understood. Most electroactive polymers are conjugated; the most commonly studied are poly(acetylene), poly(aniline), poly(pyrole), poly(thiophene), poly(phenylene sulphide), poly(phenylene) and poly(phenylene vinylene). Samples prepared by conventional chemical synthetic methods are insulating and conductivity is

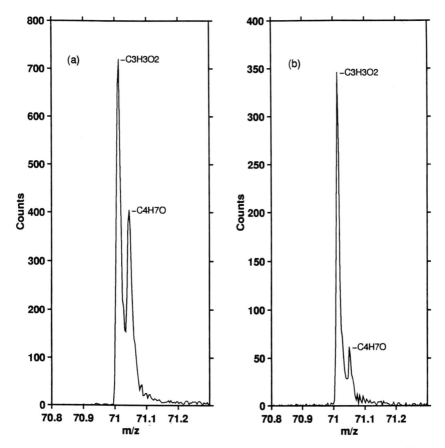

Figure 6.26 High resolution negative ion ToF SIMS spectra at *m/z* 71 for PMMA films cast from (a) THF and (b) 2-butanone. Most of the intensity in the $C_4H_7O^-$ component in (a) is due to residual THF in the film (representing the [M–H]⁻ ion, since THF is C_4H_8O) (Leeson *et al.*, 1997).

induced by doping. A preferred route, because it leads directly to films of controlled thickness grafted onto electrode supports (and which are often 'cleaner' and structurally more uniform), is electrochemical polymerisation of the monomer. An added advantage is that the prepared film is in the doped form; a polycationic structure in which counter ions taken up from the electrolyte solution provide the necessary charge balance. These films can be electrochemically switched between the conducting (doped) and insulating (undoped) state and they have promising potential in such fields as energy storage (lightweight batteries) and distribution, signal processing devices, photovoltaic cells, electronic displays, electrochemical switches, chemical and bio-sensors.

All of the information levels of XPS described in Chapter 3 are important in the study of electroactive polymers. It is not overstating the case to assert that XPS is unique in this field of polymer characterisation despite its sampling

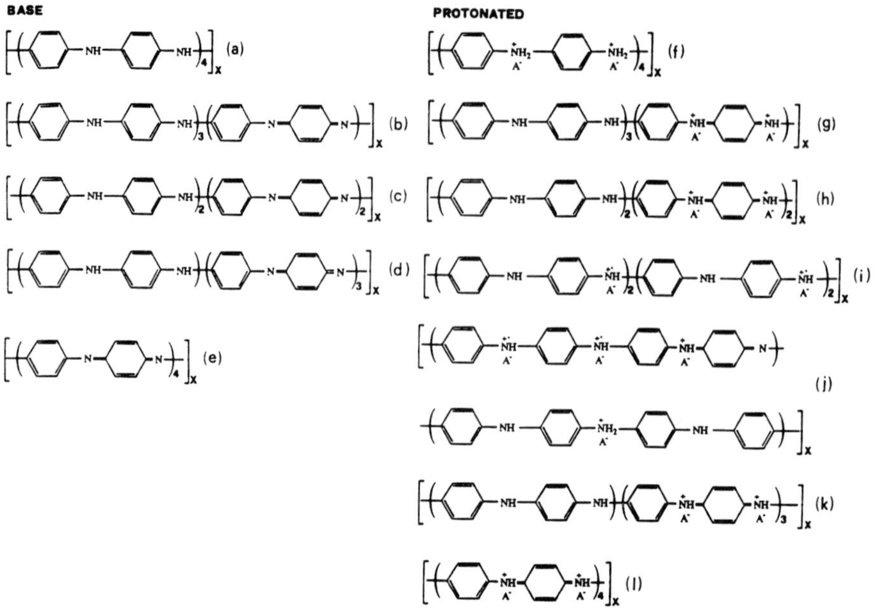

Figure 6.27 Main base and protonated forms of poly(aniline): (a), (f) leucoemeraldine; (b), (g) protoemeraldine; (c), (h), (i), (j) emeraldine; (d), (k) nigraniline; (e), (l) per-nigraniline (Malitesta *et al.*, 1993).

depth – relative to the bulk information requirements which have dominated the studies to date. This short discussion cannot begin to do justice to the large volume of literature on the subject; fortuntely there are two reviews specifically on XPS studies of electroactive polymers (Kang, Neoh & Tan, 1993; Malitesta *et al.*, 1993) and an earlier analogue which concentrates on the archetypal (but highly air-sensitive) poly(acetylene) (Salaneck, 1986). The following examples seek only to illustrate the principal areas of study.

6.6.1   Intrinsic structures

Fig. 6.27 shows the main forms of poly(aniline) in both base and protonated forms. The most reduced and the most oxidised forms are insulators, but the intermediate forms obtained either by proton addition to, or electron removal from, insulating forms can be conducting. The relationship between structure and properties is therefore complex. Fig. 6.28 shows the N 1s spectra from some of the basic forms given in Fig. 6.27, curve fitted on the basis of data from model compounds, illustrating how XPS can distinguish the $-NH-$ and $-N=$ functions. Protonated nitrogens can be further distinguished since their binding energy is $>400\,eV$ and their relative concentration can be quantitatively correlated with the concentration of anion (e.g. ClO_4^-, HSO_4^-, BF_4^-) through the core levels associated with them (Cl2p etc).

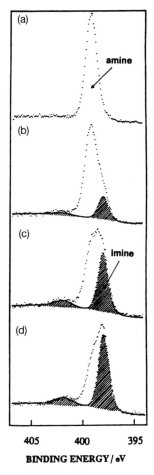

Figure 6.28 N 1s XPS spectra of poly(aniline) in the base forms: (a) leucoemeraldine; (b) protoemeraldine; (c) emeraldine, (d) nigraniline. Hatched peaks correspond to the imine (—N=) group (right) and its shake-up satellite (left), the remaining peak area is due to the amine group (Snauwaert *et al.*, 1990).

6.6.2 Doping (charge-transfer interactions)

Fig. 6.29 shows the C 1s and S 2p spectra from doped samples of poly(2,2' bithiophene) (PBT) which has the structure in the undoped form:

The S 2p peak is best fitted with two spin–orbit doublets representing neutral thiophene units and (at ~1.0 eV higher binding energy) positively polarised or partially charged sulphur species resulting from charge extraction by the dopant.

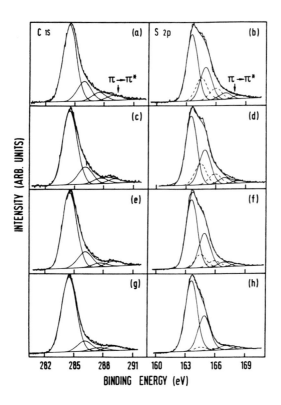

Figure 6.29 C 1s and S 2p XPS spectra of a PBT/perchlorate complex synthesised in the pressure of $Cu(ClO_4)_2 \cdot 6H_2O$ at various extents of undoping by NaOH: (a), (b) pristine sample, $ClO_4/S=0.25$; (c), (d) $ClO_4/S=0.18$; (e), (f) $ClO_4/S=0.07$, (g), (h), $ClO_4/S\sim0$ (Kang *et al.*, 1993).

The high binding energy C1s components correlate in intensity with this S2p component and represent carbon atoms in the polarised rings. Note also how the relative intensity of the shake-up satellites, probably reflecting the extended π bonding in the conducting polymers, also decreases with progressive undoping.

6.6.3   Stability and degradation

The stability of conducting polymers both under general storage conditions (reactivity to air and moisture leading to loss of desirable properties) and with respect to electrochemical cycling is a major issue. The former aspect has a significant bearing on frequent disagreements in the literature concerning XPS spectra from notionally identicial materials. Conversely, such spectral sensitivity suggests that XPS has the capability to follow degradation processes and contribute to the understanding of the resulting property variations (e.g. reduction in conductivity with time). For thin film electroactive materials it is clear that surface changes could be very important. Durability under oxidation/reduction cycling is critical to many end uses of these polymers. XPS has shown, for instance, that small concentrations of water ($<0.1\%$) in non-aqueous electrolytic solutions can have serious consequences, leading to irreversible surface chemical modification and modification of the doping/undoping process, such that anions do not leave the film during reduction and cations are driven into the polymer. For the doped system PBT–ClO_4, XPS showed that after five cycles in acetonitrile which was not perfectly dry, some of the ClO_4^- was converted to C–Cl bonds in the polymer (from the large chemical shift in the Cl2p peak and by monitoring anion exchange before and after cycling). One of the few SSIMS studies in this field confirmed this interpretation: after cycling the polymer gave a spectrum diagnostic of a chloro-substituted PBT (Morea *et al.*, 1990, 1991).

6.6.3   Electronic structure

As discussed in chapter 3 the XPS valence band probes the molecular orbitals of a polymer and provides the means to investigate the local density of filled states. When coupled with molecular orbital calculations, to help interpret the contributions of the atomic orbitals, valence band spectra have the potential to comment directly on polymer conductivity. Since the valence band photoelectrons have the highest possible kinetic energy they also sample the bulk structure to the maximum extent possible; consequently surface contamination/degradation issues are minimised. There is clearly a relationship between structures in the valence band close to the Fermi level and features in the overall core level peak shape (from discrete shake-up satellites to high binding energy asymmetry). Fundamental studies of electronic structure therefore require high sensitivity *and* high energy resolution analysis of the valence band and core level features of the polymer and related model compounds together with high quality (*ab initio*) molecular orbital calculations (see, for example, Tourillon & Jugnet (1988) and Keane *et al.* (1990)).

References

Allara, D.L., Atre, S.V. & Parikh, A.N. (1993) in *Polymer Surfaces and Interfaces II*, eds. Feast, W.J., Munro, H.S. & Richards, R.W. (Wiley, Chichester) p. 27.

Ashley, R.J., Briggs, D., Ford, K.S. & Kelly, R.S.A. (1985) in *Industrial Adhesion Problems*, eds. Brewis, D.M. & Briggs, D. (Orbital Press, Oxford) p. 199.

ASTM E42 (1991) *Surf. Interface Anal.*, **17**, 951.

Baird, R.J. & Fadley, C.S. (1977) *J. Electron Spectrosc.*, **11**, 39.

Barrie, A. (1977) in *Handbook of X-ray and Ultraviolet Photoelectron Spectroscopy*, ed. Briggs, D. (Heyden, London) p. 79.

Beamson, G. & Briggs, D. (1992a) *High Resolution XPS of Organic Polymers: The Scienta ESCA 300 Database* (Wiley, Chichester).

Beamson, G. & Briggs, D. (1992b) *Mol Phys.*, **76**, 919.

Beamson, G., Briggs, D., Davies, S.F., Fletcher, I.W., Clark, D.T., Howard, J., Gelius, U., Wannberg, B. & Balzer, P. (1990) *Surf. Interface Anal.*, **15**, 541.

Beamson, G., Clark, D.T., Kendrick, J. & Briggs, D. (1991) *J. Electron Spectrosc. Relat., Phenom.*, **57**, 79.

Benninghoven, A. (1969) *Phys. Status Solids*, **34**, K169.

Benninghoven, A. (1983) in *Ion Formation from Organic Solids* (Springer-Verlag, Berlin) p. 64.

Bletsos, I.V., Hercules, D.M., Griefendorf, D. & Benninghoven, A. (1985) *Anal. Chem.*, **57**, 2384.

Blythe, A.R., Briggs, D., Kendall, C.R., Rance, D.G. & Zichy, V.J.I. (1978) *Polymer*, **19**, 1273.

Boyd, R.D., Kenwright, A.M., Badyal, J.P.S. & Briggs, D. (1997) *Macromolecules*, **30**, 5429.

Brewis, D.M. (1982) *Surface Analysis and Pretreatment of Plastics and Metals* (Applied Science, London).

Briggs, D. (1982) *Surf. Interface Anal.*, **4**, 151.

Briggs, D. (1983) *Surf. Interface Anal.*, **5**, 113.

Briggs, D. (1987) *Org. Mass Spectrom.*, **22**, 91.

Briggs, D. (1990a) *Surf. Interface Anal.*, **15**, 734.

Briggs, D. (1990b) in *Practical Surface Analysis*, second edn, Vol 1 Auger and X-ray Photoelectron Spectroscopy, eds. Briggs, D. and Seah, M.P. (Wiley, Chichester) p. 437.

Briggs, D. (1992) in *Practical Surface Analysis*, Second Edn. Vol 2 *Ion and Neutral Spectroscopy* eds. Briggs, D. & Seah, M.P. (Wiley, Chichester) p. 367.

Briggs, D. (1993) *Spectroscopy Europe*, **5(2)**, 8.

Briggs, D. (1997) *Surf. Interface Anal.*, in press.

Briggs, D. & Beamson, G. (1992) *Anal. Chem.*, **64**, 1729.

Briggs, D. & Beamson, G. (1993) *Anal. Chem.*, **65**, 1517.

Briggs, D. & Davies, M.C. (1997) *Surf. Interface Anal.*, **25**, 725.

Briggs, D. & Fletcher, I.W. (1997) *Surf. Interface Anal.*, **25**, 167.

Briggs, D. & Hearn, M.J. (1985) *Spectrochim. Acta*, **40B**, 707.

Briggs, D. & Hearn, M.J. (1986) *Vacuum*, **36**, 1005.

Briggs, D. & Hearn, M.J. (1988) *Surf. Interface Anal.*, **13**, 181.

Briggs, D. & Kendall, C.R. (1982) *Int. J. Adhesion, Adhesives*, **2**, 13.

Briggs, D. & Ratner, R.D. (1988) *Polym. Commun.*, **29**, 6.

Briggs, D. & Wootton, A.B. (1982) *Surf. Interface Anal.*, **4**, 109.

Briggs, D., Brown, A. & Vickerman, J.C. (1989) *Handbook of Static Secondary Ion Mass Spectrometry (SIMS)* (Wiley, Chichester).

Briggs, D., Fletcher, I.W., Reichlmaier, S., Agulo-Sanchez, J.L. & Short, R.D. (1996) *Surf. Interface Anal.*, **24**, 419.

Briggs, D., Hearn, M.J., Fletcher, I.W., Waugh, A.R. & McIntosh, B.J. (1990) *Surf. Interface Anal.*, **15**, 62.

Briggs, D., Kendall, C.R., Blythe, A.R. & Wootton, A.B. (1983) *Polymer*, **24**, 47.

Brinen, J.S., Rosati, L., Chakel, J. & Lindley, P. (1993) *Surf. Interface Anal.*, **13**, 1055.

Brown, A., van den Berg, J.A. & Vickerman, J.C. (1985) *Spectrochim. Acta*, **63**, 561.

Chan, C.-M. (1993) *Polymer Surface Modification and Characterisation*, (Hanser, Munich).

Chaney, R.L. (1987) *Surf. Interface Anal.*, **10**, 36.

Chilkoti, A. (1996) in *The Wiley Static SIMS Library* (Wiley, Chichester).

Chilkoti, A. & Ratner, B.D. (1991) *Surf. Interface Anal.*, **17**, 567.

Chilkoti, A., Caster, D.G. & Ratner, B.D. (1991) *Appl. Spectrosc.*, **45**, 209.

Chilkoti, A., Ratner, B.D. & Briggs, D. (1991) *Chem. Mater.*, **3**, 51.

Chilkoti, A., Ratner, B.D. & Briggs, D. (1992) *Surf. Interface Anal.*, **18**, 604.

Chilkoti, A., Ratner, B.D. & Briggs, D. (1993) *Anal. Chem.*, **65**, 1736.

Christie, A.B. (1989) in *Methods of Surface Analysis*, ed. Walls, M.J. (Cambridge University Press, Cambridge) p. 127.

Chtaib, M., Ghijsen, J., Pireaux, J.J., Caudano, R., Johnson, R., Orti, E. & Bredas, J.L. (1991) *Phys. Rev. B*, **44**, 10 815.

Clark, D.T. (1973) in *Electron Emission Spectroscopy*, eds. Dekeyser, W. & Reidel, D. (Reidel, Dordecht).

Clark, D.T. & Kilcast, D. (1971) *Nature*, **233**, 77.

Clark, D.T., Kilcast, D., Feast, W.J. & Musgrave, W.K.R. (1973) *J. Poly. Sci., Poly. Chem.*, **11**, 389.

Cooks, R.G. & Busch, K.L. (1983) *Int. J. Mass Spec. Ion Process*, **53**, 111.

Coxon, P.A. & McIntosh B.J. (1996) Patent 744 617.

Coxon, P., Krizek, J., Humpherson, M. & Wardell, I.R.M. (1990) *J. Electron Spectrosc. Relat. Phenom.*, **52**, 821.

Davies, M.C., Lynn, R.A.P., Davis, S.S., Hearn, J., Watts, J.F., Vickerman, J.C. & Johnson, D. (1993) *J Colloid Interface Sci.*, **156**, 229.

Delcorte, A. & Bertrand, P. (1996a) *Nucl. Instr. and Meth. B.*, **115**, 246.

Delcorte, A. & Bertrand, P. (1996b) *Nucl. Instr. and Meth. B.*, **117**, 235.

Delcorte, A., Segda, B.G. & Bertrand, P. (1997) *Surf. Sci.*, in press.

Delcorte, A., Bertrand, P., Arys, X., Jonas, A., Wischerhoff, E., Mayer, B. & Laschewsky, A. (1996) *Surf. Sci.*, **366**, 149.

De Matteis, C.I., Davies, M.C., Leadly, S., Jackson, D.E., Beamson, G., Briggs, D., Heller, J. & Franson, N.M. (1993) *J. Electron Spectrosc. Relat. Phenom.*, **63**, 221.

Dilks, A. (1981) in *Electron Spectroscopy. Theory, Techniques and Applications*, Vol 4, eds. Brundel, C.R. and Baker, A.D. (Academic Press, New York) p. 277.

Drummond, I.W. (1992) *Microscopy and Analysis*, March issue, 29.

Fowler, D.E., Johnson, R.D., Van Leyen, D. & Benninghoven, A. (1990) *Anal. Chem.*, **62**, 2088.

Galuska, A. (1996) *Surf. interface Anal.*, **24**, 380.

Gardella, J.A. & Hercules, D.M. (1980). *Anal. Chem.*, **52**, 226.

Gelius, U. (1974) *J. Electron Spectrosc. Relat. Phenom.*, **5**, 985.

Gelius, U., Asplund, L., Basilier, E., Hedman, S., Helenlund, K. & Siegbahn, K. (1984) *Nucl. Instrum. Methods Phys. Res, B1*, **229**, 85.

Gelius, U., Svensson, S., Siegbahn, H., Basilier, E., Faxalv, A. & Siegbahn, K. (1974) *Chem. Phys. Lett.*, **28**, 1.

Gelius, U., Wannberg, B., Baltzer, P., Fellner-Feldeg, H., Carlsson, G., Johansson, C.-G., Larsson, J., Munger, P. & Vegerfors, G. (1990) *J. Electron Spectrosc. Relat. Phenom.*, **52**, 747.

Gilmore, I. & Seah, M.P. (1996) *Surf. Interface Anal.*, **24**, 746.

Hagenhoff, B., Lub, J., van Velzen, N.T., van Leyen, D. & Benninghoven, A. (1988) *Surf. Interface Anal.*, **12**, 53.

Hearn, M.J. & Briggs, D. (1988) *Surf. Interface Anal.*, **11**, 198.

Hearn, M.J., Briggs, D., Yoon, S.C. & Ratner, B.D. (1987) *Surf. Interface Anal.*, **10**, 384.

Hearn, M.J., Ratner, B.D. & Briggs, D. (1988) *Macromol.*, **21**, 2950.

Hicks, P.J., Daniel, S., Wallbank, B. & Comer, J. (1980) *J Phys E: Scientific Instruments*, **13**, 713.

Ishitani, T. & Shimizu, R. (1974) *Phys. Lett.*, **46A**, 487.

Jablonski, A. (1989) *Surf. Interface Anal.*, **14**, 659.

Jablonski, A. & Powell, C.J. (1993) *Surf. Interface Anal.*, **20**, 771.

Jackson, S.T. & Short, R.D. (1992) *J. Mater, Chem.*, **2**, 259.

Jede, R., Ganschow, O. & Kaiser, U. (1992) in *Practical Surface Analysis*, second edn. vol 2: *Ion and Neutral Spectroscopy*, eds. Briggs, D. & Seah, M.P. (Wiley, Chichester) p. 19.

Kang, E.T., Neoh, K.G. & Tan, K.L. (1993) *Adv. Polym. Sci.*, **106**, 135.

Keane, M.P., Lunell, S., Naves de Brito, A., Carlsson-Gothe, N., Svensson, S., Wannberg, B. & Karlsson, L. (1991) *J. Electron Spectrosc. Relat. Phenom.*, **56**, 313.

Keane, M.P., Svensson, S., Naves de Brito, A., Correia, N., Lunnell, S., Sjogren, B., Ingara, O. & Salanek, W.R. (1990) *J. Chem. Phys.*, **93**, 6357.

Kelly, M.A. & Tyler, C.E. (1972) *Hewlett-Packard J.*, **24**, 2.

Kunze, K., Stamm, M., Hartshorne, M. & Affrossman, S. (1996) *Acta Polymer*, **47**, 234.

Larson, P.E. & Palmberg, P.W. (1995) US Patent 5444242.

Leeson, A.M., Alexander, M.R., Short, R.D., Briggs, D. & Hearn, M.J. (1997) *Surf. Interface Anal.*, **25**, 261.

Leggett, G.J. & Vickerman, J.C. (1992) *Int. J. Mass Spectrom. Ion Processes*, **122**, 281.

Leggett, G.J., Briggs, D. & Vickerman, J.C. (1990) *J. Chem. Soc. Faraday Trans.*, **86**, 1863.

Linton, R.W., Mawn, M.P., Belu, A.M., DeSimone, J.M., Hunt, M.O. Jr, Menceloglu, Y.Z., Cramer, H.G. & Benninghoven, A. (1993) *Surf. Interface Anal.*, **20**, 991.

Lub, J. & van Veltzen, P.N.T. (1989) in *Ion Formation from Organic Solids, IFOSS VI*, ed. Benninghoven, A. (Wiley, Chichester).

Lub, J. & van Veltzen, F.C.B.M. (1989) *J. Polym. Sci. A. Polym. Chem.*, **27**, 4035.

Malitesta, C., Morea, G., Sabbatini, L. & Zambonin, P.G. (1993) in *Surface Characterisation of Advanced Polymers*, eds Sabbatini, L. & Zambonin, P.G. (VCH, Weinheim), p. 181.

Mamyrin, B.A., Karataev, V.I., Schmikk, D.V. & Zagulin, V.A. (1973) *Sov. Phys. JETP* (English translation) **37**, 45.

McLafferty, F.W. & Turecek, F. (1993) *Interpretation of Mass Spectra*, fourth edn. (University Science Books, Mill Valley, CA).

Morea, G., Malitesta, C., Sabbatini, L. & Zambonin, P.G. (1990) *J. Chem. Soc. Faraday Trans.*, **86**, 3769.

Morea, G., Sabbatini, L., Zambonin, P.G., Swift, A.J., West, R.H. & Vickerman, J.C. (1991) *Macromolecules*, **24**, 3630.

Moulder, J.F., Stickle, W.F., Sobel, P.E., Bomber, K.D. & Chastain, J. (1992) *Handbook of X-ray Photoelectron Spectroscopy* (Perkin-Elmer Corporation, Eden Prairie) p. 18.

Nordfors, D., Nilsson, A., Markensson, N., Svensson, S., Gelius, U. & Lunell, S. (1988) *J. Chem. Phys.*, **88**, 2630.

Ochiello, E., Morra, M., Garbassi, F., Humphrey, P. & Vickerman, J.C. (1990) in *SIMS VII* eds. Benninghoven, A. *et al.* (Wiley, Chichester) p. 789.

Oetjen, G.H. & Poschenrieder, W.P. (1975) *Int. J. Mass Spectrom. Ion. Phys.* **16**, 353.

Orchard, A.F. (1977) in *Handbook of X-ray and Ultraviolet Photoelectron Spectroscopy*, ed. Briggs, D. (Heyden, London) p. 49.

Owens, D.K. (1975) *J. Appl. Polymer Sci.*, **19**, 265.

Pireaux, J.J., Riga, J., Caudano, R. & Verbist, J. (1981) in *Photon, Electron and Ion Probes of Polymer Structure and Properties*, eds Dwight, D.W., Fabish, T.J. & Thomas, H.R., ACS Symposium Series 162 (American Chemical Society, Washington DC) p. 169.

Ratner, B.D., Yoon, S.C. & Mates, N.B. (1987) in *Polymer Surfaces and Interfaces*, ed. Feast, W.J. & Munro, H.S. (Wiley, Chichester) p. 231.

Reichlmaier, S., Bryan, S.R. & Briggs, D. (1995) *J. Vac. Sci. Technol*, **A13**, 1217.

Reichlmaier, S., Hammond, J.S., Hearn, M.J. & Briggs, D. (1994) *Surf. Interface Anal.*, **21**, 739.

Reihs, K., Voetz, M., Rundeau, F., Wolany, D. & Benninghoven, A. (1997) in *SIMS X*, eds. Benninghoven, A. *et al.* (Wiley, Chichester), p. 115.

Reilman, R.F., Msezane, A. & Manson, S.T. (1976) *J. Electron Spectrosc. Relat. Phenom.*, **8**, 389.

Riviere, J.C. (1990a) in *Practical Surface Analysis*, second edn, Vol 1: *Auger and X-ray Photoelectron Spectroscopy*, eds Briggs, D. and Seah, M.P. (Wiley, Chichester) p. 19.

Riviere, J.C. (1990b) *Surface Analysis Techniques* (Oxford Science Publications, Oxford) p. 265.

Roberts, R.F., Allara, D.L., Pryde, C.A., Buchanan, D.N.E. & Hobbins, N.D. (1980) *Surf. Interface Anal.*, **2**, 5.

Salaneck, W.R. (1986) in *Handbook of Conducting Polymers Vol II*. ed. Skotheim, T. (Marcel Dekker, New York) p. 1337.

Schofield, J.H. (1976) *J. Electron Spectrosc. Relat. Phenom.*, **8**, 129.

Schueler, B., Sander, P. & Reed, D.A. (1990) *Vacuum*, **41**, 1661.

Schweiters, J. *et al.* (1991) *J. Vac. Sci. Technol*, **A9**, 2804.

Seah, M.P. (1989) in *Methods of Surface Analysis*, ed. Walls, M.J. (Cambridge University Press, Cambridge) p. 57.

Seah, M.P. (1990) in *Practical Surface Analysis*, second edn, Vol 1: *Auger and X-ray Photoelectron Spectrocopy*, eds. Briggs, D. and Seah, M.P. (Wiley, Chichester) p. 201.

Seah, M.P. (1993) *Surf. Interface Anal.*, **20**, 243.

Seah, M.P. and Dench, W.A. (1979) *Surf. Interface Anal.*, **1**, 2.

Seah, M.P. and Smith, G.C. (1988) *Surf. Interface Anal.*, **11**, 69.

Seah, M.P. and Smith, G.C. (1990) in *Practical Surface Analysis*, second edn, Vol 1: *Auger and X-ray Photoelectron Spectroscopy*, eds. Briggs, D. & Seah, M.P. (Wiley, Chichester) p. 531.

Shard, A.G., Davies, M.C. & Schacht, E. (1996) *Surf. Interface Anal.*, **24**, 787.

Sherwood, P.M.A. (1990) in *Practical Surface Analysis*, second edn, Vol 1: *Auger and X-ray Photoelectron Spectroscopy*, eds. Briggs, D. & Seah, M.P. (Wiley, Chichester) p. 555.

Siegbahn, K., Nordling, C., Fahlman, A., Nordberg, R., Hamrin, K., Hedman, J., Johansson, G., Berkmark, T., Karlson, S.-E., Lindgren, I. & Lindberg, B. (1967) *ESCA: Atomic Molecular and Solid State Structure Studied by Means of Electron Spectroscopy* (Almquist & Wiksells, Uppsala).

Sigmund, P. (1981) in *Sputtering by Particle Bombardment 1* (Springer-Verlag, Berlin) p. 9.

Snauwaert, P., Lazzaroni, R., Riga, J., Verbist, J.J. & Gonbeau, D. (1990) *J. Chem. Phys.*, **92**, 2187.

Stradal, M. & Goring, D.A.I. (1975) *Can. J. Chem. Engng.*, **53**, 427 (and refs therein).

SurfaceSpectra (1997) The Static SIMS Library (SurfaceSpectra, Manchester).

Tanuma, S., Powell, C.J. & Penn, D.R. (1994) *Surf. Interface Anal.*, **21**, 165.

Tielsch, B.J. & Fulghum, J.E. (1994) *Surf. Interface Anal.*, **21**, 621.

Tougaard, S. & Jansson, C. (1993) *Surf. Interface Anal.*, **20**, 1013.

Tourillon, G.J. & Jugnet, Y. (1988) *J. Chem. Phys.*, **89**, 1905.

Tyler, B.J., Ratner, B.D., Castner, D.G. & Briggs, D. (1992) *J. Biomed. Mater. Res.*, **26**, 273.

Vickerman, J.C. (1989a) in *Secondary Ion Mass Spectrometry*, eds. Vickerman, J.C., Brown, A. & Reed, N.M. (Oxford Science Publications, Oxford) p. 295.

Vickerman, J.C. (1989b) in *Secondary Ion Mass Spectrometry*, eds. Vickerman, J.C., Brown, A. & Reed, N.M. (Oxford Science Publications, Oxford) p. 35.

Wagner, C.D., Davis, L.E., Zeller, M.V., Taylor, J.A., Raymond, R.H. & Gale, L.H. (1981) *Surf. Interface Anal.*, **3**, 211.

Walsh, D.J., Higgins, J.S. & Maconnachie, A. (eds) (1985) *Polymer Blends and Mixtures* (Martinus Nijhoff, Boston).

Wilding, J.R., Melia, C.D., Short, R.D., Davies, M.C. & Brown, A. (1990) *J. Appl. Polym. Sci.*, **39**, 1827.

Wittmaack, K. (1992) in *Practical Surface Analysis*, second edn, Vol 2: *Ion and Neutral Spectroscopy*, eds. Briggs, D. & Seah, M.P. (Wiley, Chichester) p. 105.

Xie, Y. & Sherwood, P.M.A. (1993) *Chem. Mater.*, **5**, 1012.

Index